AF514598

Health Informatics

This series is directed to healthcare professionals leading the transformation of healthcare by using information and knowledge. For over 20 years, Health Informatics has offered a broad range of titles: some address specific professions such as nursing, medicine, and health administration; others cover special areas of practice such as trauma and radiology; still other books in the series focus on interdisciplinary issues, such as the computer based patient record, electronic health records, and networked healthcare systems. Editors and authors, eminent experts in their fields, offer their accounts of innovations in health informatics. Increasingly, these accounts go beyond hardware and software to address the role of information in influencing the transformation of healthcare delivery systems around the world. The series also increasingly focuses on the users of the information and systems: the organizational, behavioral, and societal changes that accompany the diffusion of information technology in health services environments.

Developments in healthcare delivery are constant; in recent years, bioinformatics has emerged as a new field in health informatics to support emerging and ongoing developments in molecular biology. At the same time, further evolution of the field of health informatics is reflected in the introduction of concepts at the macro or health systems delivery level with major national initiatives related to electronic health records (EHR), data standards, and public health informatics.

These changes will continue to shape health services in the twenty-first century. By making full and creative use of the technology to tame data and to transform information, Health Informatics will foster the development and use of new knowledge in healthcare.

More information about this series at http://www.springer.com/series/1114

Arthur André
Editor

Digital Medicine

Springer

Editor
Arthur André
Pitié-Salpêtrière Hospital
Paris
France

ISSN 1431-1917 ISSN 2197-3741 (electronic)
Health Informatics
ISBN 978-3-319-98215-1 ISBN 978-3-319-98216-8 (eBook)
https://doi.org/10.1007/978-3-319-98216-8

Library of Congress Control Number: 2018960382

© Springer Nature Switzerland AG 2019
This work is subject to copyright. All rights are reserved by the Publisher, whether the whole or part of the material is concerned, specifically the rights of translation, reprinting, reuse of illustrations, recitation, broadcasting, reproduction on microfilms or in any other physical way, and transmission or information storage and retrieval, electronic adaptation, computer software, or by similar or dissimilar methodology now known or hereafter developed.
The use of general descriptive names, registered names, trademarks, service marks, etc. in this publication does not imply, even in the absence of a specific statement, that such names are exempt from the relevant protective laws and regulations and therefore free for general use.
The publisher, the authors, and the editors are safe to assume that the advice and information in this book are believed to be true and accurate at the date of publication. Neither the publisher nor the authors or the editors give a warranty, express or implied, with respect to the material contained herein or for any errors or omissions that may have been made. The publisher remains neutral with regard to jurisdictional claims in published maps and institutional affiliations.

This Springer imprint is published by the registered company Springer Nature Switzerland AG
The registered company address is: Gewerbestrasse 11, 6330 Cham, Switzerland

Contents

Chapter 1
The Information Technology Revolution in Health Care

Arthur André

1.1 "Claudes" in the Clouds

In 1865, Claude Bernard published his *Introduction to Experimental Medicine*. He exposed all the principles of modern medical science. Basically, any hypothesis has to be verified by repeated experiments to be a valid theory [1]. Improved with statistics and modern tools, these principles are still valid nowadays. In 1948, another "Claude", Claude Shannon, a Bell Company engineer, published *A Mathematical Theory of Communication* [2], which was the funding work for the information technology (IT), which will grow largely later in the century. He explained there how to quantify information from a message where the goal is to send it over a noisy channel, and then to have the receiver reconstruct the message with low probability of error, in spite of the channel noise, depending on the amount of uncertainty or entropy.

The meeting of these two "Claudes" and of these two worlds finally occurred. Information theory and its applications (computers, algorithms, artificial intelligence) have profoundly modified fields of our industry, services, and daily life. Medical science, which requires a validate proof before any practical use, now appears as changing field due to its invasion by IT.

1.2 What Is e-Health?

At the end of the 1990s, two separate—and relatively confidential—domains, at the crossroads of health care and technology, led to the new area of "e-health." The first was telemedicine, as defined as remote care delivery [3], and the second was health

A. André (✉)
Pitié-Salpêtrière Hospital, Assistance Publique Hôpitaux de Paris, Paris, France

Sorbonne Université, Paris, France
e-mail: arthur.andre@neurochirurgie.fr

© Springer Nature Switzerland AG 2019

A. André (ed.), *Digital Medicine*, Health Informatics,
https://doi.org/10.1007/978-3-319-98216-8_1

informatics, defined as the use of programmable software for health care [4]. The term "e-health" was first officially used at the Seventh International Congress on Telemedicine and Telecare in London, in November 1999, by John Mitchell from Sidney. He argued that "cost-effectiveness of telemedicine and telehealth improves considerably when they are part of an integrated use of telecommunications and information technology in the health sector." Then, he defined "e-health" as "a new term needed to describe the combined use of electronic communication and information technology in the health sector (...)."

However, the term was probably first used by tech industry and marketing people rather than academics [3]. They were inspired by other e-domains (where "e-" stands for electronic) such as e-commerce, e-business, and e-solutions that describe the transformation of many economic areas by the application of information technologies. This was undoubtedly done in an attempt to export the promises and excitement around e-business to the health-care area, to justify new investments. Intel, for example, referred to e-health as "a concerted effort undertaken by leaders in health care and hi-tech industries to fully harness the benefits available through convergence of the Internet and health care" [5].

It is interesting to note that e-health was then "theorized" by health policy researchers from developed countries where telemedicine was already developed due to population dispersion among long distances: Australia and Canada, essentially. Moreover, the aim to provide a better health-care service through these innovative techniques was already on their minds. Thus, Gunther Eysenbach from the University of Toronto delivered a speech titled "Global health equity – Medical progress & quality of life in the XXIst century" at UNESCO (Paris) in June 2001, Conference of the International Council for Global Health Progress, where he described e-health: "e-health is an emerging field in the intersection of medical informatics, public health and business, referring to health services and information delivered or enhanced through the Internet and related technologies. In a broader sense, the term characterizes not only a technical development, but also a state-of-mind, a way of thinking, an attitude, and a commitment for networked, global thinking, to improve health care locally, regionally, and worldwide by using information and communication technology" [5].

Ten years, later, however, this promise remains largely unfulfilled, as expressed by Estonian President Toomas Hendrik Ilves, Chair of the independent high-level eHealth Task Force at the European Commission in 2012: "We know that in healthcare we lag at least 10 years behind virtually every other area in the implementation of IT solutions. We know from a wide range of other services that information technology applications can radically revolutionize and improve the way..."

The lack of clear definition of this field probably comes from the initial idealistic project of merging health issues, public policy, regulatory affairs, and business in an identical and ambitious common project. For example, investors look for investments that can produce high returns even after several years. From this point of view, the specific term "telemedicine" seems inadequate, as it identifies a market niche, while e-health, as any e-thing, seems more open and promising [6]. Moreover, standard venture capital thinking hardly combines with long-term health policy, even if one goal of a reinvented heath delivery system is to achieve the "Triple Aim," the watchwords of reform driving changes in public- and private-sector actions, outlined in *Health Affairs* by Donald Berwick, Thomas Nolan, and John Whittington: better care, better health, and reduced per capita costs [7].

1.3 Institutional Definitions

These broad descriptions are reflected in the attempts to define e-health by international and government authorities.

According to the World Health Organization (WHO), e-health is a broad concept, defined as the use of electronic means to deliver information, resources, and services related to health. Many terms are included in e-health, such as:

- Electronic health records.
- Mobile health or m-health (e.g., apps, wearable technologies, medical devices).
- Telehealth or telemedicine (e.g., whereby a patient can consult a health-care worker on the computer, a tablet, or a phone).
- Health-related e-learning (use of technology and media for training and educating both a broader audience and the health workforce).
- Social media for health (informal, social online communication channels).
- Health data analysis and "big data" (transformation of data to provide insights and evidence for decision- and policymaking).

In the USA, the Food and Drug Administration (FDA) chose to define digital health as the following:

The broad scope of digital health includes categories such as mobile health (m-health), health information technology (IT), wearable devices, telehealth and telemedicine, and personalized medicine.

Providers and other stakeholders are using digital health in their efforts to:

- Reduce inefficiencies.
- Improve access.
- Reduce costs.
- Increase quality.
- Make medicine more personalized for patients [8].

For the European Commission: eHealth is the use of IT in health products, services and processes combined with organisational change in healthcare systems and new skills, in order to improve health of citizens, efficiency and productivity in healthcare delivery, and the economic and social value of health. "eHealth covers the interaction between patients and health-service providers, institution-to-institution transmission of data, or peer-to-peer communication between patients and/or health professionals" [9].

1.4 Connected Health

The wide notion of "connected health," used as an umbrella term, led to the confusion over the definitions of *telemedicine*, *telehealth*, and *m-health* [10]. Telemedicine and telehealth, relatively similar terms, have been described before. Mobile health (m-health) is defined as the practice of medicine supported by portable diagnostic devices [11]. Use of these devices at the point of care is resulting in a change in the method of health-care delivery from one that was health-systems generated to one that is remote and patient generated [12, 13]. However, connected health does not encompass other important applications of IT in medical care: health-care data collecting and analysis, robotics, and a part of personalized medicine (genomics). Therefore, we would rather use the terms *e-health*, or *digital health*, defining the whole field, and *digital medicine*, defining the part of this field that is strictly medical care.

From a practical point of view, frontiers are not so clear. As specified by Elenko et al., "digital health is a widely used term that encompasses an enormous variety of products from consumer-focused mobile apps with no clinical validation to FDA-approved apps aimed at patients, physicians or clinical pathologists to tools targeted at researchers [14]."

1.5 A New Paradigm in Health Care

The emerging information technologies allow basically the manipulation of data with an algorithm. Computers can store information as well as transform it. In a top-down approach, starting from existing technologies, one can imagine the IT applications to medicine: the aim is to find the right new technology corresponding to the actual health-care issue. We can record medical data, generate medical knowledge, help therapeutics processes, or communicate with patients [15]. Therefore, Hatcher et al. expose obvious applications of IT in the future of health care: diagnosis, treatment assignment, follow-ups, and prevention. We can add peripheral value that is not specific to health care, enhanced by IT: booking system, planning, e-learning, social networks, and wearables.

Another way of thinking is a more bottom-up approach, redefining biology through the spectrum of IT. Compared to physics, the field of biology has continuously been difficult to modelize with mathematics [18]. The remarkable complexity and variability of biology escape from purely logical descriptions, and only statistics would approach a satisfactory description. This is no longer the case with IT. A large amount of data, provided by medical records, physiological or pathological parameters, genomics, or modern imaging, could now be processed by dedicated algorithm, from simple expert systems to deep learning (see Chap. 3).

The informational view of biology, that has been largely advocated by Leroy Hood and colleagues, considers biological data that could be classified and organized according to IT concepts: (1) the digital information of the genome and the environmental signals that come from outside the genome; (2) these two *informations* are integrated together to give phenotypes, through biological *networks*; (3) biological *data* are hierarchical and multiscale across all levels of biological organization representing hierarchy of information [19]. This approach leads to a "systems" approach to disease: the application of systems biology to the challenge of human disease. The altered dynamics of information flow explain the pathophysiology of the disease and suggest new approaches to diagnosis and therapy [16]. The computational abilities to analyze large amounts of health data for each individual provide a new precision medicine paradigm. This may have two major consequences in the pursuit of modern care:

- Diagnosis and therapies may not only be driven by statistical data among a more or less homogeneous population (where individuals are similar but not identical) but also by personalized management according to each individual's genotype, biological, imaging, clinical, or environmental characteristics.
- Medicine may progressively change from reactive to proactive.

The adjunction of the growing number of "quantified-self" applications and patients' empowerment has led to the concept of "4P" medicine that is predictive, preventive, personalized, and participatory. We will deal with precision medicine and its methods—the omics—in Chaps. 5 and 6.

1.6 New Players in the Game

Since the first founding medical acts in antiquity, the doctor has always been a man of both art and science, who have expanded his knowledge and skills over the centuries and have remained at the heart of the care system. Delivering care according to the ethical criteria dictated by Hippocrates (460–370 BC), the doctor also plays the role of a researcher and an informant to patients and the rest of the general public. Progress in chemistry, physics, and microbiology in the nineteenth century, then in cellular and molecular biology, pharmacology, and imaging in the twentieth century, has transformed methods of examinations and care, assuring the scientific value of medical practice.

The growing fields of competence and action of modern medicine have fueled an asymmetry of information and power in favor of the medical profession. At the same time, compliance with ethical and medico-legal rules has become all the more important to preserve any conflict of interest.

Since the beginning of the twenty-first century, and the emergence of information technologies, we have seen the fast-growing involvement of "non-medical" actors, in particular digital companies, by providing computing resources, data science, and artificial intelligence, as well as people in general by having access to medical knowledge, thanks to these companies.

These new health players, well known to the general public, are primarily the digital giants, in particular Google, Apple, Facebook, and Amazon (GAFA) to which we can add Microsoft and IBM (GAFAMI) but also thousands of start-ups in the field that currently tend to raise awareness through modern technologies—practical and adapting to everyday life. The health-care industry is then profoundly evolving, and we can describe several technological positionings contributing to this revolution:

- MedTech, or medical technologies, i.e., the use of IT tools (software, database, apps) to answer a medical problem.
- Big data and artificial intelligence, which is derived from the first domain but specifically uses complex algorithm to generate smart care.
- Devices and robotics.
- Biotechnologies: molecular biology and genetics.
- Virtual reality, simulation.

These technologies cross several application domains in that field:

- Medical assistance for diagnosis and treatment.
- Surgical assistance and robotics.
- Electronic health records and patient management.
- Telemedicine, monitoring, and remote follow-up.
- E-learning.

All this activity around the care system generates as much hope in access to care and personalized medicine as worry about data management and the commercialization of health care.

Indeed, these recent developments in new technologies and the integration of new players in the field of digital medicine require a decision on the legal and ethical aspects different from those of conventional medicine.

1.7 Patient Empowerment, Data, and Privacy

The patient becomes the main actor of his health. This upheaval imposed by computerization offers autonomy and knowledge to the patient, not only on his health but also on what he must do to improve it. The computer revolution of the medical system is also that of the patient. Indeed, it offers him an ability to self-manage his health condition and to perform self-medication. This speed gain of the patient is closely related to the loss of speed of the doctor in his advisory role. The patient has access to a body of medical information on the Internet, more or less reliable, and can learn about the symptoms felt in isolation or in groups. More than this, he can now use a set of metric analyses at his fingertips to know his blood pressure, blood sugar, and other vital constants, which are essential in medical care, but still require analysis by competent medical skills.

The patient will be confronted with the free flow of his medical data under an anonymity that could be quickly pierced. This medical overhang is the main fear of patients jostling for the entrance of mass computerization. It is for this reason that American authorities including the National Archives and Records Administration

(NARA) and the Department of Health and Human Services (HHS) have ruled on rules for the preservation of patient privacy. The principle of this approach is to ensure true protection of patients' medical data by law (Health Insurance Portability and Accountability Act) [17]. Similarly, the European Commission recently enforced the General Data Protection Regulation (GPDR) to ensure personal data privacy. These subjects will be detailed in Chap. 9.

References

1. Bernard C. Introduction à l'étude de la médecine expérimentale. Paris, J. B: Baillière; 1865.
2. Shannon CE. A mathematical theory of communication. Bell Syst Tech J. 1948;26(3). https://doi.org/10.1002/j.1538-7305.1948.tb01338.x.
3. Della Mea V. What is e-health (2): the death of telemedicine? J Med Internet Res. 2001;3(2).
4. Lincoln TL, Builder C. Global healthcare and the flux of technology. Stud Health Technol Inform. 1998;52(Pt 2):1213–7. [published Online First: 1999/06/29].
5. Eysenbach G, Eysenbach G. What is e-health? J Med Internet Res. 2001;3(2):e20. https://doi.org/10.2196/jmir.3.2.e20. http://www.jmirorg/2001/2/e20/.
6. Rosen E. Enhancing e-health with DSL and IP video. Telemed Today. 2000;8(3):14–5. [published Online First: 2001/02/24].
7. Berwick DM, Nolan TW, Whittington J. The triple aim: care, health, and cost. Health Aff. 2008;27(3):759–69.
8. Administration USFaD. Digital health. 2017. https://www.fda.gov/medicaldevices/digitalhealth/.
9. Communication from the Commission to the European Parliament, the Council, the European Economic and Social Committee and the Committee of the Regions. eHealth action plan 2012–2020 – innovative healthcare for the 21st century. Brussels; 2012.
10. Iglehart JK. Connected health: emerging disruptive technologies. Health Aff. 2014;33:190. https://doi.org/10.1377/hlthaff.2014.0042.
11. Bhavnani SP. Scripps health and the scripps clinic division of cardiology LJ, CA, USA, Narula J, et al. mobile technology and the digitization of healthcare. Eur Heart J. 2017;37(18):1428–38. https://doi.org/10.1093/eurheartj/ehv770.
12. Silva BM, Rodrigues JJ, de la Torre Diez I, et al. Mobile-health: a review of current state in 2015. J Biomed Inform. 2015;56:265–72. https://doi.org/10.1016/j.jbi.2015.06.003. [published Online First: 2015/06/15].
13. Kumar S, Nilsen WJ, Abernethy A, et al. Mobile health technology evaluation: the mHealth evidence workshop. Am J Prev Med. 2013;45(2):228–36. https://doi.org/10.1016/j.amepre.2013.03.017. [published Online First: 2013/07/23].
14. Elenko E, Underwood L, Zohar D. Defining digital medicine. Nat Biotechnol. 2015;33(5):456–61. https://doi.org/10.1038/nbt.3222.
15. Shaffer DW, Kigin CM, Kaput JJ, et al. What is digital medicine? Stud Health Technol Inform. 2002;80:195–204. [published Online First: 2002/05/25].
16. del Sol A, Balling R, Hood L, et al. Diseases as network perturbations. Curr Opin Biotechnol. 2010;21(4):566–71. https://doi.org/10.1016/j.copbio.2010.07.010. [published Online First: 2010/08/17].
17. HIPAA Administrative Simplification. American Medical Association. 2018. https://www.ama-assn.org/practice-management/hipaaadministrative-simplification.
18. Berry G. Hyperpuissance de l'informatique. Paris: Éditions Odile Jacob; 2018.
19. Hood L, Balling R, Auffray C. Revolutionizing medicine in the 21st century through systems approaches. Biotechnol J. 2012;7(8):992–1001.

Chapter 2
Telemedicine

Pierre Simon

Clinical telemedicine and informative telemedicine were defined by the WHO in 1998 as follows [1]: "Clinical telemedicine is a professional activity that implements digital telecommunication facilities for physicians and other members of the medical profession to remotely perform medical procedures for patients." On the contrary, informative telemedicine is "an interactive audiovisual communication service that organizes the dissemination of medical knowledge and protocols for patient care and care in order to support and improve medical activity." In other words, clinical telemedicine refers to the practice of distance medicine through technological means, while informative telemedicine concerns the dissemination of knowledge and information for medical use through these same means. With more than 40 definitions of e-health in 2004, WHO took the initiative to propose in 2005 a consensual definition of e-health or cyber health [2] as "the use of IT in support health and related areas, including healthcare services, health surveillance, health literature and health education, knowledge and research," with information technology (IT) itself as "the combination of products and services that capture, store and display data and information electronically."

2.1 Informative Telemedicine

Informative telemedicine is defined as services that allow the remote delivery of medical information in order to improve medical knowledge and companies in charge of patients. This is the form of telemedicine developed in Northern Europe

P. Simon (✉)
French Society of Telemedicine, Paris, France
e-mail: pierre.simon@sf-telemed.org

© Springer Nature Switzerland AG 2019
A. André (ed.), *Digital Medicine*, Health Informatics,
https://doi.org/10.1007/978-3-319-98216-8_2

and North America, driven by the digital industry in the field of health and in legal terms of free access for members of the European Union (EU) to information society services [3]. Its applications are the most widely used and shared with the electronic medical record (or EMR) informative tele-imaging and telemonitoring services for patients with chronic diseases.

2.1.1 The French Example of DMP

The national EMR in France, the *Dossier Médical Partagé et Personnalisé* or DMP, is one of the components of a networked information system. It was launched by the health authorities in the early 2000s with the aim of sharing it among the various health professionals involved in the care of patients with chronic diseases. The March 4, 2002, Law on the rights of patients [4] states that the DMP concerns the development of diagnostic follow-ups, treatments, but also more generally all written exchanges between health professionals. The DMP is therefore made up of nominative administrative and medical information that forms a database in the sense that it is a "collection of works, data, or other independent elements, arranged in a systematic or methodically and individually accessible by electronic means or by any other means." The law on the rights of the patients [4] offers the possibility, to any person who asks for it, to have access to all information concerning his health, whether this information is held by professionals or health establishments. The patient can thus directly access his medical file; he may also choose to have access to this information through a doctor he designates for this purpose.

The Superior Council of Health Information Systems (*Conseil Supérieur des Systèmes d'Information de Santé* or CSSIS), the authority responsible for assessing the prospects and methods of implementing the medical file and the exchange of information necessary for the best care of people, considers it preferable to use the term "health record" as it is less restrictive than "medical record" in the sense that it encompasses the care of all health professionals, from physician to nurse and other regulated health professions. This recommendation was not retained and in France currently we refer to personal and shared medical records (*Dossier Médical Personalisé et Partagé* or DMP). This is the term now used in the law of modernization of the health system of January 26, 2016 [5].

Operational throughout the French territory in September 2018, it took almost 20 years for the DMP to be set up. The objectives set by the CSSIS in 1999 are still topical: facilitate the coordination of care between health professionals, the DMP must allow shared care of the patient within the different care structures of a network and sectors care, facilitate daily professional exercise by providing classification tools to retrieve information quickly according to several criteria—by nature of data (clinical, biological, imaging), in chronological order, by name, by age, by place of residence, by type of condition—and provide decision support, evaluation, and clinical studies by allowing the use of predefined protocols of care established from the reference of practices. These protocols include data entry forms corresponding to the structured data needed to assess the quality of care provided in the network, regional cooperative clinical research, epidemiological studies, and the traceability of the patient's journey through the health-care system.

The DMP accessible on the dedicated web platform is secure. In addition to these services, it offers the possibility for the patient to access his file anywhere in the world. The European directive on cross-border care of 2011 [6] allows all European citizens of the 28 member countries of the European Union (EU) to have access to their medical records from anywhere in the EU. In addition, the DMP makes it possible to promote awareness and self-management of the patient's health through the implementation of automatic alert messages (reminder of mandatory vaccinations, annual consultations or examinations, and complementary activities to be carried out).

No other European country has implemented an EMR at a level of more than 60 million inhabitants. Scotland has set up an equivalent on a scale of five to six million patients, just like Andalusia in Spain. The Nordic countries have established a national EMR, but Norway has five million individuals, Sweden nine million, and Denmark five million. It seems that the size of five to six million is an optimum, above which the success of a national EMR becomes random. The United Kingdom and the United States have only partially succeeded in setting up a national EMR, after spending a lot of money.

Let us mention the success in France of the Pharmaceutical Record (*Dossier pharmaceutique* or DP). It lists, for each beneficiary of health insurance who wishes it, all the medicines delivered during the last 4 months, whether they are prescribed by the doctor or advised by the pharmacist (21 years for vaccines, 3 years for biological drugs). The DP was created by the law of January 30, 2007, on the organization of certain health professions. Its implementation has been entrusted to the National Council of the Order of Pharmacists. Originally the DP was a patient record (DP-Patient) that allowed pharmacists only to better secure the dispensing of drugs by limiting the risk of drug interactions and redundant treatments. It is now accessible to pharmacists and doctors practicing in health-care facilities. DP improves the coordination between health professionals and the decompartmentalization of city hospitals and promotes the improvement of immunization coverage. More than 20 million DPs are now open. In addition to securing the patient, it also secures the pharmaceutical supply chain by offering other services necessary for the proper management of drug stocks: DP-Ruptures for the management of information on supply disruptions, DP-Alerts for health alerts, DP-Reminders for recalls and withdrawals of batches of drugs, and DP-Health monitoring to help monitor the health situation in France.

For the compulsory French health insurance, which was charged in 2016 by the public authorities to finalize the implementation of the DMP [5], the message sent to patients on the DMP website is that of a personal health record: "The DMP allows authorized health professionals to access information relevant to your care and to share with other health professionals medical information about you: your history, your possible allergies, the medicines you take, your reports of hospitalization and consultation, your results of exams (radios, biological analyzes ...), etc. It is a real health book always accessible and secure. To be more practical, it is computerized and you control access. Apart from you, only authorized health professionals (doctor, nurse, pharmacist ...) can consult it."

The practice of clinical telemedicine needs an EMR. The difficulties in its implementation over the last 20 years have contributed to the delay in the development of clinical telemedicine in France after the launch of the national strategy in June 2011, whereas in Northern European countries, telemedicine was able to develop, thanks to the operational nature of the national EMR. For example, Sweden was able to authorize teleconsultation in 2016 to improve access to care for isolated patients in rural areas. In France, teleconsultation funding from September 2018 will coincide with the generalization of the DMP.

2.1.2 Informative Tele-imaging

Informative teleradiology was not clinical at the beginning and does not correspond at the moment to the good practices for radiologists. It consists of sending images via a transfer platform to a teleradiologist for a diagnostic remote interpretation, with little or no clinical information. The teleradiologist who practices on these platforms does not have the possibility to ask the requesting physician to obtain additional clinical information if he considers it necessary or to propose another type of imaging more appropriate to the question asked. Initiated in the United States to improve the permanence of radiology care (telediagnosis carried out at the radiologist's home at the request of the emergency physician), this practice has evolved into tele-interpretation of scheduled radiological examinations, then to outsourcing this interpretation to international companies in order to reduce costs. It has been called low-cost teleradiology in France and considered by the radiological societies of several countries as a "dangerous assimilation of remote diagnosis to a banal service delivery." To respond to these practices, in December 2014, the Professional Council of French Radiology updated the Clinical Teleradiology Charter, which recalls the criteria for good practice [7]. This charter is based on the French Code of Medical Ethics [8], on the recommendations of the High Authority for Health (HAS) and on the telemedicine decree of October 19, 2010 [9]. For learned societies of radiology, low-cost informative teleradiology does not respect the charter of clinical teleradiology [7].

This practice of informative tele-imaging is also developing in other medical specialties (dermatology, cardiology, ophthalmology, digestive endoscopy) with the help of algorithmic solutions of artificial intelligence (AI). AI applied to medical imaging will undoubtedly be an aid to medical diagnosis. The solutions proposed must be reliable, that is to say, of a sensitivity and specificity at least equal, if not superior, to human interpretation. These solutions must be validated by scientific studies published in peer-reviewed international journals.

In addition, the way the algorithm has been built must be transparent. The algorithmic solutions of AI must today make it possible to sort between a normal image and an abnormal image, that is to say, the capacity of the algorithmic solution to give a negative result toward a suspected anomaly, that is, a 100% specificity. The diagnostic sensitivity to an abnormal image will require human intervention for a long time. Deep learning solutions can advance the sensitivity of a diagnosis, such as screening for melanoma or diabetic retinopathy [10, 11]. In summary, informative tele-imaging will be able to rely more and more on AI to exclude an anomaly on condition that the specificity is 100%. This is the criterion generally used by the international health authorities that authorize the marketing of AI solutions (FDA, CE marking).

2.1.3 Remote Monitoring Services for Patients with Chronic Diseases

Another application of informative telemedicine is home telemonitoring services for patients with chronic diseases. This form of telemedicine is the major challenge of the transformation of health systems in the twenty-first century. The goal of these new health-care organizations is to prevent aggravations of chronic diseases and hospitalization. These new organizations in France combine clinical telemedicine and informative telemedicine.

Clinical telemedicine and informative telemedicine are tested in the French national program Telemedicine Experiments for the Improvement of Health Care Pathways (*Expérimentation de Télémédecine pour l'Amélioration des Parcours En Santé* (ETAPES)) [12, 13], which is today the only telemedicine practice that brings together in a single organization the telemedicine services as defined in European law on information society services [3] and professional practices of clinical telemedicine as defined in French health law [9]. This is a financing experiment that will be evaluated in 2021 and presented to the French Parliament during the discussion of the Social Security Financing Law for the year 2022 (LFSS 2022).

The development of telemedicine services to monitor patients with chronic diseases is called for in all developed countries by the digital health industries. It is an important market creating many jobs. In a white paper published in April 2013, the two main French unions in the digital and medical device industries proposed economic models to the public authorities for the industrial development of projects championing remote surveillance of chronic diseases by 2020. To achieve an objective of at least one million patients benefiting from remote monitoring by a medical device, a working group emanating from the Strategic Committee of Health Industries (CSIS) and the Strategic Committee of the Health Sector (CSF) aired in April 2015 a report aimed at removing the obstacles to the development of these services [14]. The Agency for Shared Information Systems in Health (*Agence des systèmes d'information partagés en santé ASIP Santé*), the General Directorate of

Care Supply (*Direction Générale de l'Offre de Soins or DGOS*), the HAS, and the National Agency for the Safety of Medicines and Health Products (*Agence nationale de sécurité du médicament et des produits de santé or ANSM*) have, subsequent to this report and at the request of the CSIS, developed pedagogical fact sheets to assist in the qualification of a telemedicine project, under the leadership of the Delegation for the Strategy of Health Information Systems (*Délégation à la stratégie des systèmes d'information de santé* or DSISS).

This effort of clarification and simplification is laudable, but it has not removed all ambiguities. The French public authorities consider the services of informative telemedicine as an industrial strategy and therefore of e-commerce and European law for industrial competition, while clinical telemedicine is a professional medical practice which depends of French public health law [3]. The French Telemedicine Decree of October 19, 2010 [9], only concerns clinical telemedicine activities and their implementation conditions. EU member states have jurisdiction over their public health policy, and telemedicine is part of it in many countries. Informational telemedicine services should take into account the clinical practices of telemedicine and the national telemedicine programs. This is the experience in France in the ETAPES program until 2022 [13]. The regulatory constraints of clinical telemedicine should not be a hindrance to the development of informative telemedicine, especially home-based remote surveillance services for chronic diseases.

A second ambiguity concerns the role that health professionals must play in the national strategy expected by the digital health industries. Medical health professionals believe that the most successful chronic telemonitoring activities to date are those where the informative telemedicine service has taken into account the practices and organizations of health professionals. Projects initiated by health industry alone, without co-construction with medical health professionals, run the risk of failing.

Two French examples and two European examples argue in favor of this analysis: in France, the program of tele-observance in sleep apnea and telemonitoring of patients with an implanted defibrillator for severe heart rhythm disorders; in Europe, the Whole Systems Demonstrator (WSD) and Portavita programs.

The sleep apnea tele-observance program was aimed at deploying positive pressure night-time breathing assistance machines to 800,000 patients who needed to be connected so that the health provider could inform the compulsory health insurance of cases of non-compliance (use of the device less than 3 h a day). This purely industrial program was quickly interrupted after an appeal to the Council of State of the representatives of the patients who did not accept the "flicage" of the compulsory health insurance. It was revived when the pulmonologists, who are in charge of these patients, were able to ensure with medical providers a medical management of non-compliance in the interest of patients.

The tele-cardiology program "remote monitoring of patients with an implanted defibrillator" has had an exemplary development, as it has been co-constructed with cardiologists. They regularly made improvements to the medical device and carried out medico-economic clinical studies, which led to the validation of the defibrillator implanted and connected by the HAS.

The Whole Systems Demonstrator (WSD) program, conducted in the United Kingdom in 2009 on nearly 6000 patients in three counties, ultimately failed after being considered the largest industrial study in chronic disease telemonitoring. It was essentially aimed at demonstrating the cost-effectiveness of telemonitoring of patients with three chronic diseases (heart failure, COPD, and diabetes) [15]. Its failure is linked to methodological errors that could have been prevented if primary care and social service professionals had been involved in the construction of the program [16]. In the analysis of the impact of WSD on organizations, the authors explain that if each of the three sites had, before WSD, social care actors who could intervene in a telecare type program, then in only one site, the Health workers were trained in the telehealth program. None of the sites had a comprehensive approach to health care and social care prior to the study. The authors conclude that WSD is a complex organizational innovation that requires responsiveness and adaptation to existing local organizations, which is not possible with a controlled, randomized study. They state that "if the National Health Service (NHS) wants to develop this integrated distance care system in the future, it must first resolve the tensions observed during the WSD study among healthcare professionals and social professionals." [17].

The second European example concerns the Portavita program conducted in the Netherlands since 2002. This program, which is widely used today, illustrates the tripod on which any project for remote monitoring of chronic diseases must be based: the manufacturer who provides the services, the professionals who choose the most appropriate practices, and the patients who adhere to the solutions proposed by their treating physicians. For its construction, this program is based on several findings issued by Dutch health professionals: care protocols are not followed sufficiently, cooperation between health professionals is insufficient, and patients must play a central role in the implementation of the care process and in the analysis of its results. The company Portavita develops and offers Disease Management Systems (DMS) solutions available on the web. It does not position itself as a telemedicine operator, but only as a solution provider, and therefore does not have access to medical information. DMS solutions provide patients and health-care professionals with a standardized web interface that is fully compatible with care protocols. They make it possible to record and consult the medical data of the patient. The added value of the digital logbook is of several types: it allows the care of the patient by health professionals and the exchange of information between them through an integrated secure messaging service; the patient has access to his medical information or can even seize it (self-management); the access rights can be parameterized as well as the functions to adapt the tool to the medical organization; the organization can be driven through billing and schedule management features; and elements on financial aspects and the quality of monitoring can be collected. Portavita's DMS solutions comply with international interoperability standards. Medical data is hosted in the Netherlands by KPN, the leading Dutch telecom operator. The main telemonitoring solutions developed by Portavita concern patients with vascular thrombosis (2002), diabetic patients (2006), patients with respiratory failure (2009), and patients with heart failure (2010). Portavita is a prime example of the European model of informative telemedicine service for patients with chronic diseases. The success of this platform lies in the co-construction of solutions with general practitioners and hospital doctors. In 2013, nearly 100,000 diabetic patients were followed in the Netherlands by Care Groups (GP groups) with the Diabetes Monitoring Solution. It can be estimated that about 25% of Dutch GPs use it for the treatment of their diabetic patients. The telemonitoring solution for patients with vascular thrombosis is used by 34 anticoagulation centers (out of 65 in total) and more than 15,000 patients, i.e., almost 50% of the Dutch self-management market (the patient adjusts his treatment by himself according to a protocol proposed by his doctor). More than 4000 health professionals use it to follow more than 70,000 patients [1].

2.2 Clinical Telemedicine

The practice of medicine is an art, that is to say, a way of healing linked for a large part to the training received from practicing masters who have built the exercise of this art on the acquired knowledge of medical science and their professional experience. Clinical teaching to future doctors is delivered to the patient's bed. It is the peculiarity of French medicine to have created in the nineteenth century, in the Napoleonic era, the concept of clinical medicine, which has spread over the past two centuries in many countries around the world. Not all countries benefit from this clinical training. In countries where the doctor's degree is obtained only after a university course of 5–6 years, new general practitioners or specialists must find placements, usually in hospitals, to apply the theoretical knowledge they have acquired, before exercising their art.

In the twentieth century, the practice of medical art is illustrated, among other things, by medical consultation. The code of medical ethics, which in France has a regulatory value (decree in Council of State transcribed in the code of public health) [8], specifies the duties of doctors toward patients and doctors between them. Certain articles of the code of ethics are retained by the judges when they study the field of the responsibilities of the victim of a medical accident. The legal and regulatory obligation of the physician to provide fair, clear, and appropriate information about his condition to the person he examines, treats, or advises establishes ethical medical responsibility for the occurrence of a medical condition or medical accident. The judge will seek to know in the conclusions of the forensic report whether the victim of the medical accident was well informed of the benefits and risks of the medical practice in question, that is to say, to have been able to choose and give consent to this practice. Evidence of the information and consent given must be provided by the physician as part of the contractual responsibility that binds the physician to his patient. The physician must also demonstrate that he practices his art according to the acquired data of medical science. The judicial expert at the mission has to say whether the medical practice that generated the accident was consistent or not with the data acquired from medical science at the time of the occurrence of the accident. Finally, and this is a more recent regulatory obligation, physicians must demonstrate that they maintain and improve their knowledge in their obligation of continuous professional development. Since the code of medical ethics also applies to this form of remote medical practice [18], any physician who decides to practice clinical telemedicine must respect it.

France and the many countries that have applied the model of clinical medicine have defined telemedicine as a form of remote medical practice that uses IT. The scope of practice of telemedicine is defined in France by law and an implementing decree for the conditions of implementation.

The French law Hospital, Patients, Health and Territory (HPST) of July 21, 2009 [19], specifies in its article 78 that telemedicine is a form of remote medical practice using the technologies of information and communication. It connects, among themselves or with a patient, one or more health professionals, including necessarily a medical professional and, where appropriate, other professionals providing care to the patient. It makes it possible to establish a diagnosis; to ensure, for a patient at risk, a preventive follow-up or a post-therapeutic follow-up; to request a specialized opinion; to prepare a therapeutic decision; to prescribe products; and to prescribe or to perform services or acts or monitor the condition of patients.

The implementing decree of October 19, 2010 [9], gives the definitions of five telemedicine practices that can be summarized as follows: teleconsultation when a patient remotely consults a medical health professional; teleexpertise when two health professional medical experts or more are giving away their expert advice on a patient's medical file; medical remote monitoring when a patient with a chronic disease is followed at home by clinical and/or biological indicators selected by a patient; medical health professional advice, spontaneously collected by a medical device equipped with an AI algorithm or seized by the patient or a medical assistant, then transmitted to the medical professional via informative telemedicine services; medical remote assistance when a medical professional attends remotely to a non-medical health professional; and medical response as part of the medical regulation that comes in several medical services including urgent medical assistance and personalized medical e-consultancy (*médical teleconseil*). It follows several regulatory conditions for implementation, including the prior consent of the patient to this remote medical practice and the obligation to trace in the patient's medical file or the DMP the conditions for carrying out the act and the incidents and possible techniques that have occurred. France and the others countries that have applied the model of clinical medicine have defined clinical telemedicine as a form of remote medical practice that uses ICTs. In France, the framework for the clinical practice of telemedicine is defined by the law and an decree for the conditions of implementation.

2.2.1 Teleconsultation

Teleconsultation may have several definitions depending on whether the country considers it as a clinical form of remote medical practice. It is thus necessary to distinguish between teleconsultation platforms that are informative telemedicine services and teleconsultation medical practices that are part of clinical telemedicine.

Two types of teleconsultation can be distinguished: the one programmed by the doctor and the immediate or unscheduled one requested by a person. The latter is often a personal medical e-consultancy [20].

2.2.1.1 Scheduled Teleconsultation

In France, teleconsultation must take into account patients' rights to prior information so that consent can be obtained. This is the reason for recommending teleconsultation by the High Authority for Health (*Haute Autorité de Santé* (HAS)) [21]. Thus, to respect the great ethical and deontological principles, a teleconsultation cannot be done without a prior information of the persons concerned, in a fair, clear, and appropriate way, in order to collect their consent to this new medical practice [9]. The teleconsultation can only be programmed, that is to say, it cannot replace a first consultation face to face, unless the interest of the patient justifies it. The programming of a teleconsultation makes it possible to provide access to the patient's computerized medical file or the DMP if it has been opened by the patient. The file is essential for the doctor who performs a teleconsultation, especially in a person with chronic diseases that alternates face-to-face consultations and teleconsultations.

The practice of teleconsultation is not always well understood in relation to the traditional exercise of the medical art. This is a classic, face-to-face consultation that is conducted remotely by audiovisual communication means [13]. When a doctor performs a classic consultation, he meets a person who considers himself sick. If it is a first appointment, it spends more time than when it is a follow-up consultation for a chronic disease. The first consultation necessarily involves a clinical examination, a prolonged interrogation on the personal and family antecedents, the treatments received, the surgeries carried out, etc. It includes an intellectual act that allows the physician to fully understand the subject of the complaint or request and to assess the health status of the new patient. Can a first teleconsultation do as well? It is doubtful, because the climate of confidence that develops between the patient and his doctor during a first meeting is based on the warmth of a direct face-to-face relationship that cannot be reproduced by a first-time remote consultation by videoconference. In addition, it is exceptional that a first consultation does not require a clinical examination. All these points justify that the first consultation of a patient must take place face to face. It allows to inform and to obtain its consent when a follow-up by teleconsultation is proposed to him.

If it is not recommended when the person has easy and quick access to the doctor, a first teleconsultation may however be justified in certain circumstances, especially when the interest of the patient justifies it. This is the case in the acute phase of a stroke: the neurologist on duty in the neurovascular unit carries out a teleconsultation with the emergency department where the patient was received to judge the level of disability related to stroke and of the indication of thrombolysis after consulting brain imaging. This is also the case in the overseas regions for island populations: a first teleconsultation can be beneficial to judge the need or not to perform an evacuation by air ambulance. We can also think that for the prison population, the first teleconsultation is preferable to an extraction for a hospital consultation, if the state of the prisoner does not require a clinical examination (e.g., for a first consultation of psychiatry). Finally, a person with a severe disability and bedridden in a residential care facility, or in another institution, can benefit from a first teleconsultation to avoid a difficult and tiring journey [22]. All these first consultations are done with the help of health professionals who are with the patient.

What would a patient think of a treating physician who during a face-to-face consultation would not bother to view the computerized medical record on their computer or DMP? Scheduled teleconsultation is indicated in a patient provided that the physician has access to his computerized record or to the DMP. This is a regulatory obligation for the physician [9]. Teleconsultation between two face-to-face consultations is an added value in the follow-up of a patient with one or more chronic diseases because it allows a more regular follow-up without moving the patient. For example, teleconsultation at home is preferred by renal transplant patients. It prevents them from moving to the hospital and losing a day of work.

The Organization in France

The organization of teleconsultations in France must respond to the recommendations of the HAS [23]. It modifies the medical organizations of the reference health institutions that offer teleconsultation and teleexpertise to small local institutions without medical specialists. Larger health facilities have developed advanced consultations in small institutions since the 1990s. Today, teleconsultation can optimize medical time in referral facilities by reducing the time when physicians provide outpatient consultations. In outpatient medicine, it can also optimize the medical time devoted to follow-up of Elderly clinical (*Etablissement d'Hébergement pour Personnes Agées Dépendantes* (EHPAD)) patients [22]. The patient must consent to this new form of care after having been informed of the benefits and risks of telemedicine [23]. Consent should not be collected at each teleconsultation act if it is performed regularly, if it is part of the personalized care plan, and if the patient has initially agreed to this new form of care. However, he must be informed that he can return to usual face-to-face care if he is not satisfied.

From a technical point of view, a teleconsultation must meet certain requirements [20]. First of all, the bit rate of the digital network must be sufficient for the image to be of good quality. The definition of videotransmission screens has progressed, and it is possible to make a good teleconsultation today with a digital bit rate lower than 1 Mbit/sec. The confidentiality of the exchange between the patient and his doctor must be ensured by the use of adapted helmets equipped with microphones. This is important when the teleconsultation takes place in a room where there are several patients, as is the case during the teleconsultation performed during a hemodialysis session [24]. The visual conditions for performing a teleconsultation may vary according to the uses and indications. Teleconsultation can be performed in a room dedicated to telemedicine with wall screens of greater or lesser width, the larger ones can create the atmosphere of a tele-presence [20]. These large screens are necessary when several people are attending the teleconsultation. Such equipment, however, is expensive. They are also useful for mixed function rooms: teleconsultation, teleexpertise, tele-teaching, multidisciplinary consultation meeting of recourse as performed in oncology or other medical specialties. The solution of the mobile telemedicine cart can be used when teleconsultation is performed in the patient's bed as in teledialysis or EHPAD [20]. Cabins for teleconsultation booths to collect certain clinical constants (blood pressure, heart rate, temperature, weight, cardiorespiratory rhythm, etc.) as well as imaging exams interpreted by AI (EKG, tympanic examination, examination of the eye, examination of the skin) are now proposed. The tools of health connected with AI enrich the practice of teleconsultation, which can thus approach the quality of a consultation face to face. These cabins can be installed in isolated areas, in nursing homes, pharmacies, or other places suitable for this purpose [20]. In the context of mobile health, tablets can be used for teleconsultations, including the assistance of a nurse at the home of the person consulted, to examine chronic wounds and take pictures. Self-guided carts or robots, equipped with screens, are also experienced for teleconsultation in nursing homes or in dialysis rooms, even to carry out medical visits in health facilities. The evolution toward a robotization of teleconsultation remains ethically questionable for several reasons: the technological performance gives the impression to the patient to address a "robotic" doctor, which can dehumanize the relation at a distance; the contribution of the nurse to the medical act disappears; and finally, the service rendered to the patient unable to move appears secondary to the fact of avoiding the removal of the doctor [20].

As specified in the telemedicine decree [9], psychologists may be present with the patient during a teleconsultation, including psychiatry. This profession and all regulated health professions who attend a teleconsultation are required to respect medical or professional secrecy. This point is very important because these innovative solutions cause in the institutions the curiosity of the people of the care team. When they are present at the teleconsultation, the doctor must remind them of their obligation to respect the confidentiality of the exchanges.

In the spirit of the law and the decree of telemedicine [9, 19], the act of teleconsultation is carried out under the responsibility of a medical professional as defined in the code of public health. It concerns doctors, midwives, and dentists. The teleconsultation funding under the common law of the compulsory health insurance is now obtained and will come into effect in September 2018. The fee is that of a traditional consultation face to face (25 € for the general practitioner and 28 € for the specialist doctor).

2.2.1.2 Immediate Teleconsultation and the Personalized Medical e-consultancy

Due to the initiative of private organizations (insurers, complementary health, other organizations), teleconsultation platforms have been developing for a few years now. This immediate access to a doctor is increasingly appreciated by urban workers who cannot find time to see their doctor or who do not want to wait for long hours in the emergency department. These new practices of immediate teleconsultation, followed sometimes by a tele-prescription, concern mainly benign affections. Such platforms are also developing in several European countries (the United Kingdom, Sweden, Switzerland).

Is this practice safe and secure for the patient and the doctor? Opinion polls confirm the interest of these new practices since the interviewees estimate that 70% of the medical consultations can today be done by Internet without a physical examination. This possibility is of primary interest to parents with young children. If a fever or cough in an infant or young child corresponds in most cases during an epidemic to a viral condition that will be treated only by symptomatic drugs (antipyretic, analgesic), we cannot exclude a more severe condition that would have been better recognized by a physical examination. To date, there is little scientific work to validate this immediate teleconsultation. In the event of a serious medical accident, the parents would not fail to turn against the doctor by reproaching him for failing to perform a physical examination. These teleconsultation platforms are more like medical e-consultancy platforms where the platform's doctor reassures, advises, or directs according to his perception of the necessity and not of a face-to-face consultation with a physical examination.

The immediate medical e-consultancy [20] is a societal demand that has emerged for less than a decade, linked to new ways of experiencing time and prioritizing urgency and immediacy in all economic, social, and professional activities [25]. The digital revolution contributes to this evolution. The medical e-consultancy issued by telephone, during a call to the center 15, illustrates this societal evolution with a rise in the number of calls of about 5% per year for 10 years. The 15 centers received more than 31 million calls in 2016, including less than 1 million for a life-saving emergency. The doctor of the medical regulation gives several types of answers to these calls. Among them, the "medical council" in health is described by the HAS [26] as "the complement of general information in response to a request for advice, whenever the regulator considers that the call does not require in emergency medical consultation." The HAS specifies that it is "an act that can only be performed by the regulating doctor" because "it constitutes a medical prescription." Finally, the regulating physician must "record it in the medical regulation file of the patient and must specify to the caller to renew his call if he notes a persistence, a resumption or an aggravation of the symptoms."

The personalized medical e-consultancy is considered in France as a legal practice of clinical telemedicine. According to medical regulatory experts, it accounts for almost 60% of the 31 million annual calls. While vital emergency calls have dropped significantly in the last decade with the reduction of road accidents, the medical e-consultancy is progressing strongly, reflecting the change in the company's behavior toward the community's health.

Is the personalized medical e-consultancy different from an immediate teleconsultation? This subject is under debate. It is important to distinguish these two practices for the following reasons. The personalized medical e-consultancy is based on information provided by the caller. This information is very fragmentary compared to that collected during a classic medical procedure where the doctor has the medical record and a clinical examination time. The medical e-consultancy cannot therefore be assimilated to a consultation or teleconsultation. Teleconsultation is also different from the medical e-consultancy because it is a generally programmed act that has been previously approved by the patient, clearly informed of the benefits and risks of this practice [9]. In addition, teleconsultation, whether immediate or scheduled, must be done by videotransmission [13], whereas most medical teleconferencing platforms are telephones. The programmed nature of a teleconsultation, recommended by the HAS [21], gives the doctor time to have access to the medical data necessary for the realization of the act [9], that is to say, to the clinical data, biological, or radiological patient record. The personalized medical e-consultancy is not programmed, which does not allow the doctor to have sufficient and objective knowledge of the applicant's medical history. Thus, the personalized medical e-consultancy by Internet or by telephone platform cannot fulfill the ethical conditions of implementation of a teleconsultation. As recalled by the National Council of the Order of Doctors in 2009 [9], the act of telemedicine is a medical practice in its own right, and as for its indication and quality, it cannot be a degraded form.

The personalized medical e-consultancy has an interest in answering users' requests, in order to make a filter between what is an immediate medical consultation or a deferred consultation with the attending physician and what is possibly a real emergency not received by the appellant.

The practice of the personalized medical e-consultancy is not without risk for the doctor. She should not be paid on call but on vacation or salary. This form of fixed remuneration is found in all European countries that have developed the medical e-consultancy (Sweden, the United Kingdom, Switzerland, etc.). The organizer of the platform must ensure compliance with regulatory requirements. The practice of telemedicine, in general, hardly lends itself to an acute or emergency situation, as a physical examination is usually necessary. This is the reason why most of the complementary mutuals or insurance companies that manage these medical e-consultancy platforms immediately inform their members not to use them in case of emergency but to appeal to an emergency call center.

When an attending physician is solicited over the phone by a member of his patient and gives him or her advice on a symptom or its treatment, he agrees to do so because he knows the medical record of the appellant and he thinks he can provide him with advice appropriate to his condition. In such a situation, the risk of medical error is not greater than a face-to-face consultation. The practice of telephone counseling, widespread in general practice, is generally downstream of a recent consultation and is somehow part of the same care. However, it is important for the doctor to make a phone call in the patient's medical file. It is different when a doctor practices the personalized medical e-consultancy for people he does not know.

The development of AI solutions could in the short or medium term modify the analysis that has just been done on the medical e-consultancy. These platforms held today by medical professionals or advanced practice nurses could be replaced by Chatbot robots. This solution is now being tested by the NHS in the United Kingdom [27]. Similarly, with the possibility of access to the DMP of a caller, the conditions for performing immediate teleconsultation could improve and integrate into a care path with the treating physicians, especially when they are inaccessible. Finally, solutions to help medical diagnosis by AI can only strengthen the safety and reliability of these new practices.

2.2.2 *Teleexpertise*

Teleexpertise is probably the telemedicine activity that will most structure new medical organizations in the twenty-first century. In fact, no doctor can today claim to possess a comprehensive knowledge of medical science to deal with the health problems of a person as a whole. Over the years, medicine has reached a level of scientific complexity which partly explains the phenomenon of medical specialization, or even overspecialization, which has marked the last 30 years.

Teleexpertise represents a new way of practicing medicine [20]. It enables medical professionals to consult each other regularly, to pool their medical knowledge, and to enhance their reciprocal competence. It only respects the doctor's ethical duties toward patients [8]. Teleexpertise promotes the continuity of care and avoids the break that today represents too long appointment times in certain specialties. Direct teleexpertise between physicians, usually in the absence of the patient, can replace the specialized consultation and thus shorten the usual time for obtaining a specialized opinion. It allows the attending physician, coordinator of care, to ensure continuity of care in better conditions for the patient. This person must be informed of this practice and consent to it because it replaces certain specialized face-to-face consultations to which the patient, particularly the patient suffering from chronic illnesses, used to go and may wish to maintain. The patient must keep the free choice of the mode of specialized medical follow-up; that is why he must be informed of the possibility of this new practice, from which he can draw benefits in terms of continuity and quality of care. The telemedicine decree [9] states that "professionals participating in a telemedicine act may, unless the duly informed person objects, exchange information relating to that person, in particular through the use of information and communication technologies." If the patient can oppose the teleexpertise, it is because he has been informed before. Consent should not be collected for each act of teleexpertise if the patient has initially agreed to the attending physician to this new form of care. He should be informed, however, that he can return to face-to-face consultations with the medical specialist if he is not satisfied. The patient may be attached to his specialist physician and may wish to have a face-to-face consultation from time to time.

The question of the tool used to carry out the teleexpertises is asked. The choice is the responsibility of the requesting physician who must comply with the telemedicine decree [9]. If the attending general practitioner already has the practice of videoconferencing teleconsultation, the same tool can be used for programmed synchronous teleexpertise, provided that the requested specialist physician is himself equipped and available. Today, there are low-cost, computer-based videoconferencing solutions. Teleexpertise by telephone, widely practiced in recent years, is no longer recommended as it will cause failure of the requesting doctor and the doctor asked to remember to ensure the traceability of information given and received in the patient's medical file. Such negligence would be blamed on the doctor if his responsibility was implicated in a medical accident of a patient who had benefited from a teleexpertise by telephone. Realizing a teleexpertise with the same equipment and protocols devolved to teleconsultation puts the doctor away from such a risk.

Asynchronous teleexpertise by secure messaging in health (*messagerie sécurisée en santé* or MSS) is the recommended form of teleexpertise that will be the most commonly practiced in the short term. It allows the secure transfer of medical pictures, such as skin picture, EKG, etc., and confidential clinical information in writing. So, the answer of teleexpert physician can integrate the medical record ou DMP. The situations of the teleexpertise practice are numerous [20].

The teleexpertise in primary care is used to avoid the patient to move in classic consultation with the medical specialist. The attending physician directly consults his specialist colleague on the basis of elements in the medical file. This way of working is demanded by the new generation of general practitioners who, during their internship, have made the habit of consulting their hospital colleagues about their patients. Young doctors are destabilized when they arrive on the ground of the liberal exercise and they no longer have the opportunity to seek specialist advice almost immediately as in the hospital. In primary care, quickly obtaining such advice facilitates the coordination of care and avoids a break in continuity of care. The young doctors understood that telemedicine made it possible to do otherwise than to ask patients to make their own appointment with a specialist with waiting times of several months. Teleexpertise with the specialist saves time, enhances the role of the attending physician, and strengthens interprofessional cooperation. In addition, the pooling of knowledge between general practitioners and medical specialists has the advantage of mutually enhancing skills. It is the learning function of telemedicine.

The practice of a regular, even daily, teleexpertise between public or private health facilities is an inevitable organizational evolution to improve the care path of patients in a health territory or region. As demonstrated by the study of the Midi-Pyrénées region conducted in the 1990s [20], the practice of inter-institutional teleexpertise makes it possible to better manage the hospitalization of a patient and to avoid, once in two (50%), an unnecessary transfer to the referral hospital or university hospital. In addition, it corrects once in five (20%) a possible loss of luck. It allows emergency services to avoid certain hospitalizations, especially those requested for specialized advice. The teleexpertise requested by the emergency physicians of the small health establishments from the medical specialists of the reference establishment makes it possible to obtain a first specialized opinion, the installation of a therapy, the recommendation not to hospitalize the patient or to reverse the transfer, or the organization of an external consultation.

The success of a teleexpertise practice makes it necessary to review the medical organizations whatever the place of exercise. In the hospital sector, medical specialists in the reference establishment must integrate into their daily practices a time of offer of service to the small establishments of the territory.

In the outpatient sector, the demand for specialized expertise also needs to be organized differently. Multiprofessional health-care homes are the primary care structures that have most anticipated these developments. They will benefit from the organizations set up in the health establishments. It is also important that the liberal specialists organize themselves to offer primary care physicians their tele-expertise service.

Second-opinion teleexpertise was one of the first applications of telemedicine in radiology. Radiologists distinguish telediagnosis as "the exploitation of the transmission of images for the remote obtaining of a primary and definitive diagnosis, in the absence, from the patient of a radiologist to interpret these images immediately." The absence of a radiologist from the patient justifies that telediagnosis is assimilated to a teleconsultation, the teleradiologist having the possibility, if he deems it necessary, to interrogate the patient remotely. The assimilation of telediagnosis to a radiological teleconsultation has been recognized in France in the latest teleradiology charter [7]. Radiological teleexpertise differs from telediagnosis; it is a second opinion given to the requesting radiologist by a more expert radiologist. For radiology departments, the expert radiologist authorized to give a second opinion must meet at least two of the following criteria: the recognition by professionals of their organ specialty, an important daily practice in the various pathologies within its domain expertise, a minimum number of files seen per year, participation in staff and multidisciplinary consultation meetings, and possibly research and teaching activities in the field concerned. In addition, the expert radiologist must practice in relation to or belong to a center of competence or reference [7].

The permanence of teleradiology care at the level of a territory, or even of a region, is of two natures. When the establishment of a given territory does not have a radiologist to the patient, it is the radiologist of the reference institution, or even a private radiologist company, who provides a teleconsultation for the interpretation of the examination. This requires that the radiological examination rooms be equipped with videoconferencing systems to enable the teleradiologist required to interrogate the patient if necessary and that a regional or national digital platform can promote image transfers, their accommodation, and telemedicine practices between institutions.

In the field of pathology, telepathology practices between experts have developed in recent years, at the regional, national, or even international level. It is most often second-opinion teleexpertise, sometimes enriched with diagnostic interpretation by AI algorithms [28, 29]. It is in the field of telepathology that teleexpertise could be assimilated to a telediagnosis.

Teleexpertise is now remunerated in France under the common law of Social Security. This act of telemedicine is not recognized in all countries and is sometimes considered a teleconsultation. It is developing rapidly in developing countries that are beginning the epidemiological transition to chronic diseases of aging. The use of teleexpertise at a national or even international level in these countries, particularly in sub-Saharan Africa, makes it possible to compensate for the shortage of specialist doctors [30].

References

1. Simon P. Champ et enjeux de la télémédecine, 17-46, in *Télémédecine – Enjeux et pratiques*, Collection Synthèses et repères, Le Coudrier Editions, Brignais, octobre 2015, 190 p. http://www.edition-lecoudrier.fr/produit/7/9782919374083/Telemedecine%20-%20Enjeux%20et%20pratiques.
2. World Health Organization. Legal frameworks for ehealth. In: Global observatory for eHealth services, vol. 5. 2012. http://apps.who.int/iris/bitstream/handle/10665/78076/9789242503142_fre.pdf;jsessionid=164A3CC840C37EF1D7B02F3BCF37ECB1?sequence=1.
3. Directive 2000/31/CE du parlement européen et du Conseil du 8 juin 2000 relative à certains aspects juridiques des services de la société de l'information, et notamment du commerce électronique dans le marché intérieur ("directive sur le commerce électronique"). http://eur-lex.europa.eu/LexUriServ/LexUriServ.do?uri=OJ:L:2000:178:0001:0016:FR:PDF.
4. LOI n° 2002-303 du 4 mars 2002 relative aux droits des malades et à la qualité du système de santé. JORF du 5 mars 2002, p.4118. https://www.legifrance.gouv.fr/affichTexte.do?cidTexte=JORFTEXT000000227015&categorieLien=id.
5. LOI n° 2016-41 du 26 janvier 2016 de modernisation de notre système de santé. JORF du 27 janvier 2016. https://www.legifrance.gouv.fr/affichTexte.do?cidTexte=JORFTEXT000031912641&categorieLien=id.
6. Directive 2011/24/UE du Parlement européen et du Conseil du 9 mars 2011 relative à l'application des droits des patients en matière de soins de santé transfrontaliers. https://eur-lex.europa.eu/legal-content/FR/TXT/HTML/?uri=CELEX:32011L0024.
7. Société Française de Radiologie. Charte de la téléradiologie. Sept.2009. http://www.sfr-net.org/Data/upload/documents/Groupe%20de%20travail/Téléradiologie/Charte%20Téléradiologie%20G4.pdf.
8. Conseil national de l'ordre des médecins. Code de déontologie. Ed. Avril 2017. http://www.sfrnet.org/Data/upload/documents/Groupe%20de%20travail/Téléradiologie/Charte%20Téléradiologie%20G4.pdf.
9. Décret n° 2010-1229 du 19 octobre 2010 relatif à la télémédecine. JORF du 21 octobre 2010. https://www.legifrance.gouv.fr/affichTexte.do?cidTexte=JORFTEXT000022932449&categorieLien=id.
10. Esteva A, Kuprel B, Novoa RA, Ko J, Swetter SM, Blau HM, Thrun S. Dermatologist-level classification of skin cancer with deep neural networks. Nature. 2017;542(7639):115–8.
11. Rahimy E. Deep learning applications in ophthalmology. Curr Opin Ophthalmol. 2018;29(3):254–60.
12. Arrêté du 6 décembre 2016 portant cahiers des charges des expérimentations relatives à la prise en charge par télésurveillance mises en œuvre sur le fondement de l'article 36 de la loi n° 2013-1203 de financement de la sécurité sociale pour 2014. https://www.legifrance.gouv.fr/eli/arrete/2016/12/6/AFSH1633837A/jo.
13. LOI n°2017-1836 du 30 décembre 2017 de financement de la sécurité sociale pour 2018, promouvoir l'innovation en santé, article 54, JORF du 30 décembre 2017. https://www.google.fr/search?source=hp&ei=mikcW8_yEMnoUtrDvZgH&q=article+54+de+la+LFSS+2018&oq=article+54+de+la+LFSS+2018&gs_l=psy-ab.3...11673.25181.0.26767.27.26.0.0.0.0.77.1610.26.26.0....0...1.1.64.psy-ab..1.25.1530.0..0j35i39k1j0i131k1j0i22i30k1j0i22i10i30k1j33i160k1j33i21k1.0.QPfdQxmxwto.
14. GT 33 CSIS-CSF : Permettre l'émergence d'une stratégie industrielle en matière de e-santé, en soutien de la politique de santé publique, en associant les industriels. Lever les freins au déploiement de la télémédecine. Sept. 2015. http://www.snitem.fr/sites/default/files/ckeditor/rapport-gt33-csis-csf-dssis-sans-annexes.pdf.
15. Hendy J, Chrysanthaki T, Barlow J, Knapp M, Rogers A, Sanders C, Bower P, Bowen R, Fitzpatrick R, Bardsley M, Newman S. An organisational analysis of the implementation of telecare and telehealth : the whole systems demonstrator. BMC Health Serv Res. 2012;12:403–17.

16. Steventon A, Bardsley M, Billings J, Dixon J, Doll H, Hirani S, Cartwright M, Rixon L, Knapp M, Henderson C, Rogers A, Fitzpatrick R, Hendy J, Newman S, Whole System Demonstrator Evaluation Team. Effect of telehealth on use of secondary care and mortality : findings from the WSD cluster randomised trial. BMJ. 2012;344:3874–92.
17. Steventon A, Bardsley M, Billings J, Dixon J, Doll H, Beynon M, Hirani S, Cartwright M, Rixon L, Knapp M, Henderson C, Rogers A, Hendy J, Fitzpatrick R, Newman S. Effect of telecare on use of health and social care services: findings from the Whole Systems Demonstrator cluster randomised trial. Age Ageing. 2013;42(4):501–8.
18. Conseil national de l'ordre des médecins. Télémédecine. Janvier 2009. https://www.conseil-national.medecin.fr/sites/default/files/telemedecine2009.pdf.
19. LOI n° 2009-879 du 21 juillet 2009 portant réforme de l'hôpital et relative aux patients, à la santé et aux territoires. JORF du 22 juillet 2009. https://www.legifrance.gouv.fr/affichTexte.do?cidTexte=JORFTEXT000020879475&categorieLien=id.
20. Simon P. Les différents actes de télémédecine, 77-108, in *Télémédecine – Enjeux et pratiques*, Collection Synthèses et repères, Le Coudrier Editions, Brignais, octobre 2015, 190 p. http://www.edition-lecoudrier.fr/produit/7/9782919374083/Telemedecine%20-%20Enjeux%20et%20pratiques.
21. Haute autorité de santé. Qualité et sécurité des actes de téléconsultation et de téléexpertise. Avril 2018. https://www.hassante.fr/portail/upload/docs/application/pdf/2018-04/fiche_memo_qualite_et_securite_des_actes_de_teleconsultation_et_de_teleexpertise_avril_2018_2018-04-20_11-05-33_441.pdf.
22. Salles N. *Télémédecine en EHPAD, les clés pour se lancer*. Ed. Le Coudrier, 69530 Brignais ; octobre 2017. http://www.edition-lecoudrier.fr/produit/10/9782919374199/Telemedecine%20en%20EHPAD%20-%20Les%20cles%20pour%20se%20lancer.
23. Haute autorité de santé. Grille de pilotage et de sécurité d'un projet de télémédecine. Juin 2013. http://www.has-sante.fr/portail/upload/docs/application/pdf/2013-07/guide_grille_de_pilotage_et_de_securite_d_un_projet_de_telemedecine_2013-07-18_13-34-47_545.pdf.
24. Charasse C, Boulahrouz R, Leonetti F, Potier J, Stanescu C, Le Cacheux P, Ang KS, Baluta S. Teledialysis in satellite hospital: 5-year practice in Saint-Brieuc. Nephrol Ther. 2013;9(3):143–53.
25. Aubert N. L'urgence, symptôme de l'hypermodernité : de la quête de sens à la recherche de sensations. Communication et organisation. 2006;29:11–21.
26. Haute autorité de santé. Modalités de prise en charge d'un appel de demande de soins non programmés dans le cadre de la régulation médicale, mars 2011. www.has-sante.fr/portail/jcms/.../regulation-medicale-argumentaire.
27. Burguess M. The NHS is trialling an AI chatbot to answer your medical questions. Wired, 5 Jan 2017. http://www.wired.co.uk/article/babylon-nhs-chatbot-app.
28. Alami H, Fortin JP, Gagnon MP, Pollender H, Têtu B, Tanguay F. The challenges of a complex and innovative telehealth project : a qualitative evaluation of the eastern Quebec telepathology network. Int J Health Policy Manag. 2017;7(5):421–32.
29. Maxmen A. Deep learning sharpens views of cells and genes. Nature. 2018;553(7686):9–10.
30. Diby F, Zadi S, Adoubi A, Nguessan E, Ehua S. Connaissances, attitudes et pratiques des professionnels de la santé face aux opportunités de diagnostic et de prise en charge des pathologies cardiovasculaires par la télémédecine en Côte d'Ivoire. Eur Res Telemed. 2017;6:35–42.

Chapter 3
Artificial Intelligence and Health Care

Bruno Peyrou, Jean-Jacques Vignaux, and Arthur André

3.1 What Is Artificial Intelligence?

The term "artificial intelligence" (AI) was mentioned for the first time in 1956 by John McCarthy during a conference where several scientists decided to meet to see if machines could be made intelligent. Since then, AI is usually defined as the capability of a computer program to perform tasks or reasoning processes that we usually associate with intelligence in a human being. Often it has to do with the ability to make a good decision even when there is uncertainty or vagueness or too much information to handle [1].

In 1950, British mathematician and father of computer science Alan Turing in his famous paper, "Computing Machinery and Intelligence," examined the question, "Can machines think?" through the *Imitation Game* [2]: "a human interrogator is tasked with determining which of two chat-room participants is a computer, and which is a real human." The interrogator can say or ask anything, but all interaction is solely through typed text. If the interrogator cannot distinguish computer from human with a given accuracy after interacting with each participant, then the computer has passed the Turing Test. To date, no computer was able to pass. There is no consensus as to when we might reach that stage [3]. We can however consider the IBM computer system Watson's victory at the quiz show Jeopardy in 2011 and Google Duplex with an AI reservation-making assistant in 2018 as the conclusive success of machines.

Indeed the goal of artificial intelligence is to make machines less literal, thus abstracting concepts from limited experience and transferring knowledge between

B. Peyrou (✉) · J.-J. Vignaux
iTechCare Medical Data Research, Paris, France

A. André
Pitié-Salpêtrière Hospital, Assistance Publique Hôpitaux de Paris, Paris, France

Sorbonne Université, Paris, France

© Springer Nature Switzerland AG 2019
A. André (ed.), *Digital Medicine*, Health Informatics,
https://doi.org/10.1007/978-3-319-98216-8_3

domains or "generalizing." This human-level intelligence is called artificial general intelligence (AGI) or "strong AI" [4]. In comparison, "weak AI" or narrow AI (ANI) are systems designed for a specific task, also called recognition intelligence. Algorithms need to be fed with enormous amounts of data to be able to reach recognition intelligence or ANI [3]. An intermediate level is represented by cognition intelligence, where machines can infer rules from data beyond pattern recognition. This is now feasible thanks to deep learning methods.

For now, we must agree with Deo that "although there are thousands of papers applying machine learning algorithms to medical data, very few have contributed meaningfully to clinical care" [5]. Nevertheless, a November 2014 report by Ernst & Young (EY) on life sciences discusses the enormous potential of big data in the health domain. It is a real economic machine that is poised to sweep over the health systems of the world. According to this report, only groups that have anticipated the change in the economic system will be able to respond as best as possible to the evolution of the medicine and pharmacy markets. Actually, for some years now, these groups have been collecting raw data that they store waiting for legislation to evolve about the private data market for the health sector. At the top of the list of these groups is Google, which funds Flatiron Health, a company that aims to bring together all the factors that promote the positive prognosis of people with cancer. To do this, they collect and analyze data from millions of patients. Not only does Google direct its attention to the analysis of epidemiologic mass trends or genetic screening programs; we can also talk about IBM, among others, and the service that they offer through Watson for Health. This service is manifold and offers solutions to health-care actors to optimize their processes through Watson Care Manager, but also to help in the prognosis of cancer through Watson for Genomics, in the drug discovery research through Watson for Drug Discovery, as well as in the medical literature understanding through Watson for Oncology.

3.2 Expert Systems

Expert systems represent the lowest level of AI. Expert systems are comprised of a knowledge; an inference engine, which uses IF-THEN-ELSE rules to make sense of the dataset; and a user interface, where a user receives the required information [6]. Such systems are best applied to automate calculations and logical processes where rules and outcomes are relatively clear. "Expert systems work the way an ideal medical student would: they take general principles about medicine and apply them to new patients" [7]. Despite relatively simple technology behind, experts systems could be very powerful in medical assistance.

In his very complete e-book *A Guide to Artificial Intelligence in Healthcare*, Bertalan Mesko lists many of the current applications of AI in medicine [3]. Indeed, many patient-oriented medical chatbots are based on expert systems: Florence is a practical chatbot for older patients that reminds them to take their pills [8]. SafedrugBot embodies a chat messaging service that offers assistant-like support to doctors who need appropriate information about the use of drugs during breastfeeding. Izzy helps women track their period and serves as a birth control pill reminder. Efficient diagnosis assistant systems are also based on expert systems: bots like Your.Md or CitizenDoc aim to help patients find a solution to the most common symptoms through AI. However, a chatbot never replaces an experienced doctor. If the patient's issue seems severe or doubtful, the bot itself addresses the user to an appointment with a doctor for a diagnosis and eventually for the prescription of a therapy. In a specific domain concerning millions of patients with sometimes difficult access to care, some authors have described an algorithm to screen tuberculosis, based on a rule base, knowledge base, and patient database architecture. All these experts systems are based on a relatively recent subfield of computer science called machine learning.

3.3 Machine Learning Algorithms

Machine learning, according to Arthur Samuel, gives "computers the ability to learn without being explicitly programmed."

As a field of study, machine learning sits at the crossroads of computer science, statistics, and a variety of other disciplines concerned with automatic improvement over time and inference and decision-making under uncertainty [9]. The uncertainty of a diagnosis inferred from a sign could be basically seen as a mathematical problem dealing with probabilities. The application of probability theory to learning from data is called Bayesian learning [10]. An algorithm is trained to execute a task (e.g., to distinguish cancer from non-cancer tissues) and to improve its performance and accuracy with experience. Depending on the amount of data and computational resources, these processes can gain time and precision compared to human subjective judgment. Machine learning techniques often deal with a large amount of data and therefore have a more realistic approach of biological and medical problems than "simple" mathematical science and modelization.

3.4 Machine Learning Paradigms

Three different paradigms can be classically described in machine learning.

3.4.1 *Supervised Learning*

In supervised learning, the algorithm focuses on a known output; in that case the goal is to produce a probabilistic prediction y' in response to a query x'. Supervised machine learning models are trained on prelabeled data referred to as the training set of (x, y) pairs. The training error is roughly defined by the difference between predicted outcome y' and actual outcome y. Supervised learning systems form their predictions via a learned mapping $f(x)$, which produces an output y for each input x. Supervised learning includes several probability algorithms, like decision trees, decision forests, logistic regression, support vector machines, neural networks, kernel machines, and Bayesian classifiers [9]. The classical example in medicine is the automated interpretation of an ECG or a CT scan. These are all tasks that a trained physician can do and so the computer is often trying to approximate human performance [5] *and maybe surpass it*. The number of parameters of such a model is critical: more parameters reduce bias but increase noise (and random associations). The complexity of a model is a trade-off between bias and variance [9].

The medical applications of supervised learning are diagnosis assistance for rare, difficult pathologies, and also follow-up, that needs an accurate evaluation of treatment outcome: one application on children diagnosed with cerebral palsy with support vector machines shows an overall accuracy of 96.80% in diagnosis using only two easily obtainable basic gait parameters [11]. Similarly, the diagnosis of valvular heart disease through neural network ensembles (see further *Artificial Neural Networks*) has been made with an accuracy of 97.4% [12].

3.4.2 Unsupervised Learning

In contrast, in unsupervised learning, there are no outputs to predict. The goal is to find patterns or groupings within the data. No training is needed because there is no desired output. The underlying concept is called clustering, that is to say, the assumptions that among a dataset, the machine can automatically regroup features (symptoms, images, genes). A criterion function is defined as one that embodies these assumptions—often making use of general statistical principles such as maximum likelihood, the method of moments, or Bayesian integration—and optimization or sampling algorithms are developed to optimize the criterion [9].

The "fuzzy" aspect gives the algorithm the flexibility to classify a data point to each cluster to a certain degree relating to the likelihood of belonging to that cluster. This method can detect hidden patterns across large and complex datasets. Unsupervised learning can be very useful in identifying mechanisms or risk factors for complex multifactorial diseases [5]. Fuzzy c-means (FCM), an effective clustering method developed by Bezdek [2], is one of the most common instruments for medical diagnosis [13]. It has already been proposed for classification of thyroid diseases [14] or tumors by gene characterization [15]. This method also has its weaknesses: the need to determine membership cutoff, i.e., the distance from the feature to the center of the clusters, roughly represented by the mean of the considered data and the fact that clusters are formed whatever the inputs, i.e., if the data belong to no cluster at all, the algorithm will do it anyway, thus potentially leading to errors. FCM has shown its efficiency in primary headache, where diagnosis is often difficult. The algorithm showed an accuracy (proportion of true results—whether positive or negative) of 0.97 for migraine [13].

3.4.3 A Combined Method: Reinforcement Learning

Instead of training examples that indicate the correct output for a given input, the training data in reinforcement learning are assumed to provide only an indication as to whether an action is correct or not. The action taken by the algorithm is reinforced by a reward if it is correct; if the action is incorrect, the problem of finding the correct one remains.

3.5 Deep Learning

Unsupervised learning and reinforcement learning are the fundamental principles for deep learning that are applied among successive layers of the dataset. Deep learning discovers intricate structure in large datasets by using the backpropagation algorithm to indicate how a machine should change its internal parameters that are used to compute the representation in each layer from the representation in the previous layer [16]. Deep learning can therefore offer an exciting solution to incomplete biomedical knowledge: these methods can discover new biomarkers without any human input, conceivably generating truly unexpected discoveries [17].

Some algorithm results have demonstrated similar performance to human experts in the assessment of diabetic retinopathy, dermatological lesions [18], atrial fibrillation screening [19], or Parkinson's disease [20].

3.6 An Example of Machine Learning Algorithm: The Artificial Neural Network

Artificial neural networks (ANNs) are the foundation of autonomous artificial intelligence. Their conceptualization derives schematically from the functioning of the brain and neurons, in particular the encoding phase by the neural engrams, i.e., the capacity of a neural network to create memory according the Hebbian principles (ref). ANNs are computer algorithms that introduce numerical and statistical learning methods that themselves work through perceived stimuli and make decisions.

The simplest neural network is the perceptron invented in 1957 by Frank Rosenblatt; it is only made of two neurons in the input layer connected via synapses to one neuron in the output layer. Now, a typical neural network has an input layer, a hidden layer, and an output layer each composed of several neurons; this can go up to several thousands of neurons in each layer and up to several tenths of layer for the deepest structures. Each neuron has an input and an output, the output is a mathematical function of the input called the activation function. And each output is linked to the next neuron (input) by a connection which is weighted via a formula including a weight and a bias. The weights and biases are changed via training by the backpropagation algorithm which retro-propagates the error calculated on the last layer between the expected answer and the one given by the ANN. The importance of the connections between neurons increases or decreases over the training and the dataset used.

For example, the neurons in the input layer could represent the health information collected during the outpatients' clinics. The hidden layer represents the pathologies that deteriorate a specific function in an individual. And the output layer can represent the outcome of such a cluster of symptoms (on the risk for the life of the patient).

In the literature and since the last decades, we find different methods using ANN such as multi-layered perceptrons, sigmoidal multi-layer neurons, convolutional neural networks, Kohonen networks, Hopfield networks, etc.

3.7 Algorithms for Decision-Making

Medical knowledge and publications are increasing exponentially. As explicitly described by B. Mesko: "in 1950, it was estimated that the doubling time for medical knowledge was about 50 years. In 1980 it was about 7 years for medical knowledge to double. In 2010, it only took about 3.5 years. It is estimated that by the year 2020, it will only take 73 days for the volume of medical knowledge to double". On MEDLINE, between 1978 and 2001, a total of 8.1 million journal articles were published. On PubMed, there are 23 million papers. In the domain of physicians trained in epidemiology, it would take an estimated 627.5 h per month to evaluate these articles [21]. The quantity of medical knowledge is simply impossible for a person to retain even if this person was reading the material all day long, whereas a computer program can read all the data, store the relevant information, and use it to help in medical decision-making. These kinds of algorithms are based on natural language processing (NLP); they can understand a text written by a person and extract intent, sentiment, or meaning.

The IBM AI program, Watson, launched its special algorithm for oncologists able to provide clinicians evidence-based treatment options. The purpose is to support tumor board meetings: "Watson for Oncology has an advanced ability to analyse the meaning and context of structured and unstructured data in clinical notes and reports that may be critical to selecting a treatment pathway, combining attributes from the patient's file with clinical expertise, external research, and literature data" [3].

Interviewed by B. Mesko, Van Oijen, clinical epidemiologist and associate professor at the Academic Medical Center, University of Amsterdam, believes that

Watson for Oncology could result in a reduction of costs and efforts. He thinks that "Watson for Oncology in its current form is perfectly designed for use in preparation for and during multidisciplinary tumour boards. It is the ideal additional discipline, providing up-to-date support during the tumour board meeting, but can also triage (prioritize) on before the order in which patients are discussed during the tumour board meetings."

Professor Dr. S.P. Somashekhar, Chairman of the Manipal Comprehensive Cancer Center in Bangalore, India, presented at the San Antonio Breast Cancer Symposium in 2016 a study evaluating Watson. Manipal Hospitals has put Watson to test, comparing the evidence-based recommendations of Watson for Oncology to that of Manipal's tumor board (the team of 12–15 cancer specialists who review the most complex cancer cases each week). In a double-blinded study, the doctors at Manipal found that Watson was concordant with the tumor board recommendations in 90% cases of breast cancer.

Some of the UK cancer specialists find Watson's utility more limited for now: "giving patients information and receiving feedback, although potential applications are legion" [22].

Watson expertise has however been questioned. The specialized scientific media STAT examined Watson for Oncology's use, marketing, and performance in hospitals across the world, from South Korea to Slovakia to South Florida. It appears that the system is till now closer to the Memorial Sloan Kettering tumor board (the expert board that helped to develop the algorithm) than to a universal up-to-date tool.

Notably, STAT says, while it has emphatically marketed Watson for cancer care, IBM hasn't published any scientific papers demonstrating how the technology affects physicians and patients [23].

Other solutions have been developed: "Microsoft's research machine-learning project, dubbed Hanover, also aims to ingest all the papers and help predict which drugs and which combinations are the most effective for cancer cases. Moreover, a series of start-up, such as the Hungarian Turbine aim to build smart algorithms to make oncology better" [3].

3.8 Algorithms in Radiology: The Era of Radiomics

Traditionally, medical imaging has been a subjective or qualitative science. Nowadays, the amount of imaging datasets combined to computational methods provides a comprehensive quantification combined to visual assessment [24]. The use of computational algorithms to analyze medical images is now called "radiomics" [17, 25, 26]. Even if the authors may not all agree with the extent of this new domain, a large definition of radiomics can include two machine learning-applied procedures: classical image analysis with human eyes-defined features which is derived from supervised learning and deep learning with feature learning, whose principle is close to an unsupervised learning [17].

A pioneer work was conducted by Ito et al. [27] concerning bone textures, leading to the idea that properties and patterns of pathological tissues (shape, intensity, texture) can be described by the radiologist, but also define imaging features which can be computerized [24].

Radiology's major potential resides in its ability to assess the characteristics of human tissue noninvasively and therefore is routinely used in clinical practice for diagnosis, treatment guidance, and monitoring.

Some quantitative measurement methods have been created and validated to overcome visual interpretation and its subjectivity. In oncology, precision medicine (for diagnosis and staging, which is linked to prognosis) involves not only genomics and proteomics but also spatial and density characterization, provided by modern imaging [26, 28]. Response evaluation criteria in solid tumors (RECIST) is a set of published rules that define cancer evolution during treatments. The criteria were originally published in February 2000 by an international collaboration including the European Organization for Research and Treatment of Cancer (EORTC), the National Cancer Institute (NCI) of the United States, and the National Cancer Institute of Canada Clinical Trials Group. An update was published in January 2009 [29]. Nowadays, the many clinical trials evaluating cancer treatments for objective response in solid tumors are using the RECIST platform, which is developed by Parexel Informatics [30]. This kind of standardized criteria, still dependent on the physician's observation, can be considered as the first step to replace the radiologist's eye by the computer. In magnetic resonance imaging (MRI), some research in prostate cancer has shown that textural features were significantly correlated with Gleason score [31, 32].

In supervised radiomics, the methodology can be resumed in four steps: image acquisition, extracting radiomics features from the images, training the computer, and validating a prognosis score. The extracting stage still needs human experts' intervention as image features still are human-defined features.

These "image features" are mathematical descriptions of the visual properties of an image, describing the low-level visual information present in an image, such as the intensity/brightness and the texture of image regions [25, 26].

Aerts et al. analyzed 440 radiomic features quantifying tumor phenotypic differences based on its image intensity, shape, and texture. In a large dataset of 1019 CT scans from lung and head-and-neck cancer patients, they found that a large number of radiomic features have prognostic power, of which many of their prognostic implications have not been described before providing a method that can quantify and monitor phenotypic changes during treatment [28].

The main challenge is to define mathematically a complex pattern or "high-level" features. In some specific applications, high-level features are amenable to mathematical definition or can be constructed by combining multiple low-level features [17]. These form the image biomarkers currently used today (e.g., counting the number of high-density pixels in the wall of an artery to quantify the atherosclerotic plaque burden).

Segmentation is one the most critical aspects in the attempts to generalize radiological diagnosis. Manual delineation is a straightforward solution but can also be very time-consuming and is susceptible to inter-observer variability [33]. Therefore semi-automatic or automatic could be a great alternative. In deep brain structure imaging [24, 34], Chupin et al. showed that radiomic features extracted from 3D Slicer volumes had a significantly higher reproducibility and were more robust than those extracted from manual segmentation [35].

A radiomic biomarker definition requires robust approaches that analyze all of the available variations in an image. Deep learning appears to be an ideal tool to accomplish this goal. Oakden et al. presented a proof of concept for the feasibility of imaging deep learning-based prognosis tools [17].

3.8.1 *Reproducibility and Repeatability*

Any quantitative imaging measurement requires both repeatability and reproducibility. Repeatability relates to the uncertainty in obtaining the same result in the same patient when he or she is examined more than once on the same system. Reproducibility relates to the ability to yield the same result for the same patient when that patient is examined on different systems and at different imaging sites [36]. Zhao et al. used an image database of 31 patients with lung cancer with same-day repeated CT scans, to assess the impact of slice thickness and reconstruction algorithm on the stability of 89 radiomic features. They concluded that repeatability of features derived from scans with the same imaging settings was good; however, only 19% of the features were repeatable when different settings were used [37]. It has been shown that up to 75% of the CT or PET radiomic features vary more than 5% with respiration, making breath-hold or four-dimensional imaging acquisition necessary for a correct assessment of the lesions [38]. MRI is an even harder challenge in terms of reproducibility mainly because of the distortion [39, 40]. PET is a quantitative imaging technique and therefore requires a common quality control (QC)/quality assurance (QA) procedure to maintain the accuracy and precision of quantitation [36]. PET scans were used to show a good test–retest stability in up to 71% of the radiomic features [41].

3.9 Health Assistance and Patient Management

Looking at primary care, doctors and nurses often meet patients with minor issues that could be treated without intervention from a medical profession, persons who only want prescription renewal and reassurance or who have organizational questions. There are plenty of start-ups and companies offering similar solutions, for example, Kore.ai, and we believe there will be many more in the future [3].

The AiCure app, supported by the National Institutes of Health in the United States and which uses AI associated with a smartphone webcam, ensures that patients are adhering to their prescriptions, coaching them in order to manage their health condition. This is very useful for people with chronic diseases, for patients who tend to go against the physicians' advice, and for participants in clinical trials [42]. Close to that, the virtual nurse Molly was developed by the medical start-up Sense.ly. The interface uses machine learning to support patients with chronic conditions in between doctors' visits [43].

The United Kingdom's National Health Service (NHS) has tested a beta version of a chatbot app, co-developed with Babylon Health, for dispensing remote medical advice, with the aim of reducing the burden on its non-emergency helplines and waiting time in the hospital emergency rooms. Babylon Health [44] apps are able to provide medical AI consultation based on personal medical history and common medical knowledge. Users report the symptoms of their illness to the app, which checks them against a database of diseases using speech recognition. After taking into account the patient's history and circumstances, Babylon offers an appropriate course of action. However this app, provided by a private sector company, has recently been removed from the NHS Digital Apps Library [45]. Results of this experiment are still pending.

3.10 Conclusion

AI has still to prove its efficiency based on scientific criteria in the medical field and convince the public of medical value. According to a YouGov survey for the British Science Association of more than 2000 people, 70% of respondents are happy with intelligent machines carrying out jobs such as crop monitoring, but this falls to a miserly 23% when talking about medical operations in hospitals [46].

Finally, machine learning does not solve any of the fundamental problems of causal inference in observational datasets. "Algorithms may be good at predicting outcomes, but predictors are not causes" [7, 47]. AI algorithms should definitely be a powerful assistance, and not a substitute, to doctors. The new machine learning methods analyze complex datasets and have the abilities to tease out subtle statistical regularities from massive datasets, and the humans have the abilities to draw on diverse background knowledge to generate plausible explanations and suggest new hypotheses [9]. From this clever collaboration, Chen says in the New England Journal of Medicine: "Yet if the future will not necessarily resemble the past, simply accumulating mass data over time has diminishing returns" [48].

References

1. Rossi F. Artificial intelligence: potential benefits and ethical considerations: European Parliament's Citizens' Rights and Constitutional Affairs Policy Department. 2016. www.europarl.europa.eu
2. Turing AM. Computing machinery and intelligence. Mind. 2018;LIX(236):433–60.

3. Mesko B. A guide to artificial intelligence in healthcare, The Medical Futurist, Editor. 2017. https://leanpub.com/ArtificialIntelligenceinHealthcare.
4. Yao M. @thinkmariya. WTF is artificial intelligence?!. TOPBOTS. 2017. https://www.topbots.com/wtf-is-artificial-intelligence-intro/.
5. Deo RC. Machine learning in medicine. Circulation. 2015;132:1920–30.
6. tutorialspoint.com. Artificial intelligence expert systems. 2018. Available from: https://www.tutorialspoint.com/artificial_intelligence/artificial_intelligence_expert_systems.htm.
7. Obermeyer Z, Emanuel EJ. Predicting the future — big data, machine learning, and clinical medicine. N Engl J Med. 2016;375:1216.
8. Florence – your health assistant. 2018. https://www.florence.chat.
9. Jordan MI, Mitchell TM. Machine learning: trends, perspectives, and prospects. Science. 2015;349(6245):255–60.
10. Ghahramani Z. Probabilistic machine learning and artificial intelligence. Nature. 2015;521:452–9.
11. Kamruzzaman J, Begg RK. Support vector machines and other pattern recognition approaches to the diagnosis of cerebral palsy gait. IEEE Trans Biomed Eng. 2006;53(12 Pt 1):2479–90.
12. Das R, Turkoglu I, Sengur A. Diagnosis of valvular heart disease through neural networks ensembles. Comput Methods Prog Biomed. 2009;93(2):185–91.
13. Wu Y, Duan H, Du S. Multiple fuzzy c-means clustering algorithm in medical diagnosis. Technol Health Care. 2015;23(Suppl 2):S519–27.
14. Azar AT, El-Said SA, Hassanien AE. Fuzzy and hard clustering analysis for thyroid disease. Comput Methods Prog Biomed. 2013;111(1):1–16.
15. Sun L, Xu J. Feature selection using mutual information based uncertainty measures for tumor classification. Biomed Mater Eng. 2014;24(1):763–70.
16. LeCun Y, Bengio Y, Hinton G. Deep learning. Nature. 2015;521:436–44.
17. Oakden-Rayner L, et al. Precision radiology: predicting longevity using feature engineering and deep learning methods in a radiomics framework. Sci Rep. 2017;7(1):1648.
18. Esteva A, et al. Dermatologist-level classification of skin cancer with deep neural networks. Nature. 2017;542:115–8.
19. Halcox JPJ, et al. Assessment of remote heart rhythm sampling using the AliveCor heart monitor to screen for atrial fibrillation: the REHEARSE-AF study. Circulation. 2017;136(19):1784–94.
20. Ma C, et al. An efficient diagnosis system for Parkinson's disease using kernel-based extreme learning machine with subtractive clustering features weighting approach. Comput Math Methods Med. 2014;2014:985789.
21. Alper BS, Hand JA, Elliott SG, Kinkade S, Hauan MJ, Onion DK, Sklar BM, et al. How much effort is needed to keep up with the literature relevant for primary care? J Med Libr Assoc. 2004;92(4):429–37.
22. Alder Hey Children's Hospital set to become UK's First 'Cognitive' Hospital – Science and Technology Facilities Council. 2018. Available from: https://stfc.ukri.org/news/alder-hey-children-s-hospital-set-to-become-uk-s-first-cognitive-hospital/.
23. Ross C. @caseymross. IBM pitched Watson as a revolution in cancer care. It's nowhere close. 2017. https://www.statnews.com/2017/09/05/watson-ibm-cancer/.
24. Larue RT, et al. Quantitative radiomics studies for tissue characterization: a review of technology and methodological procedures. Br J Radiol. 2017;90(1070):20160665.
25. Kumar V, et al. Radiomics: the process and the challenges. Magn Reson Imaging. 2012;30(9):1234–48.
26. Lambin P, et al. Radiomics: extracting more information from medical images using advanced feature analysis. Eur J Cancer. 2012;48(4):441–6.
27. Ito M, et al. Trabecular texture analysis of CT images in the relationship with spinal fracture. Radiology. 1995;194(1):55–9. https://doi.org/10.1148/radiology.194.1.7997582.
28. Aerts HJ, et al. Decoding tumour phenotype by noninvasive imaging using a quantitative radiomics approach. Nat Commun. 2014;5:4006.
29. Eisenhauer EA, et al. New response evaluation criteria in solid tumours: revised RECIST guideline (version 1.1). Eur J Cancer. 2009;45(2):228–47.

30. parexel.com. 2017. Available from: https://www.parexel.com/perceptive-close.
31. Gnep K, et al. Haralick textural features on T2-weighted MRI are associated with biochemical recurrence following radiotherapy for peripheral zone prostate cancer. J Magn Reson Imaging. 2017;45(1):103–17.
32. Wibmer A, Hricak H, Gondo T, et al. Haralick texture analysis of prostate MRI: utility for differentiating non-cancerous prostate from prostate cancer and differentiating prostate cancers with different Gleason scores. Eur Radiol. 2015;25:2840. https://doi.org/10.1007/s00330-015-3701-8.
33. Eminowicz G. McCormack M Variability of clinical target volume delineation for definitive radiotherapy in cervix cancer. Radiother Oncol. 2015;117(3):542–7. https://doi.org/10.1016/j.radonc.2015.10.007.
34. Ji Z, et al. Generalized rough fuzzy c-means algorithm for brain MR image segmentation. Comput Methods Prog Biomed. 2012;108(2):644–55.
35. Chupin M, et al. Automatic segmentation of the hippocampus and the amygdala driven by hybrid constraints: method and validation. NeuroImage. 2009;46(3):749–61.
36. Bolleard R, et al. FDG PET/CT: EANM procedure guidelines for tumour imaging: version 2.0. Eur J Nucl Med Mol Imaging. 2015;42:328–54.
37. Zhao B, et al. Reproducibility of radiomics for deciphering tumor phenotype with imaging. Sci Rep. 2016;6:23428.
38. Oliver JA, et al. Variability of image features computed from conventional and respiratory-gated PET/CT images of lung cancer. Transl Oncol. 2015;8(6):524–34.
39. Spatial precision in magnetic resonance imaging–guided radiation therapy: the role of geometric distortion. Int J Radiat Oncol Biol Phys. 2017.
40. Holl EM, et al. Improving targeting in image-guided frame-based deep brain stimulation. Neurosurgery. 2010;67(2 Suppl Operative):437–47.
41. Leijenaar RTH, et al. Stability of FDG-PET Radiomics features: An integrated analysis of test-retest and inter-observer variability. Acta Oncol. 2013;52(7):1391–7. https://doi.org/10.3109/0284186X.2013.812798.
42. AiCure. 2018. Available from: https://aicure.com/.
43. Sensely. How are you feeling today? 2018. Available from: http://www.sensely.com/.
44. NHS doctor appointments online. GP at hand. 2018.
45. @digitalhealth2. Babylon and others removed from NHS Apps Library. 2018. https://www.digitalhealth.net/2018/03/babylon-removed-nhsapps-library/.
46. British Science Association. One in three believe that the rise of artificial intelligence is a threat to humanity. 2018. tishscienceassociation.org/news/rise-of-artificial-intelligence-is-a-threat-to-humanity.
47. Kleinberg J, et al. Prediction policy problems. Am Econ Rev. 2015;105(5):491–5.
48. Chen JH, Asch SM. Machine learning and prediction in medicine — beyond the peak of inflated expectations. N Engl J Med. 2017;376(26):2507–9. http://dx.doi.org.gate2.inist.fr/10.1056/NEJMp1702071.

Chapter 4
Blockchain and Health

Cécile Monteil

Many hopes are vested in the use of Blockchain in health care, but is it a temporary fad or is it going to revolutionize the field? Blockchain is a technology and technologies are tools, not solutions. There are some friction points in health care where Blockchain can really make a difference, and others where it will be completely useless.

4.1 Blockchain and the Magic Notebook Analogy

4.1.1 The Magic Notebook

To explain the Blockchain, we can use a simple metaphor called the "magic notebook." Imagine a notebook with numerous copies that different people hold. Each copy is a mirror of each other in a way that if you write something in one copy, it instantly appears in every other copy. The magic notebook is always in sync.

Building on this metaphor, if one were to write a sentence in the notebook and cross it out later, or modify any other entry, it would be visible from every copy, and the modifications would instantly be rejected. The magic notebook keeps securely all entries.

One thing that makes the magic notebook special is that it has unlimited pages. People can write as many notes as they like, and the book will always create a new page. In addition, it is impossible to rip out a page as it would break the number sequence. Ripping out a page from your copy would automatically break the link between the pages.

C. Monteil (✉)
Pediatric Emergency Room of Robert Debré Hospital, AP-HP, Paris, France

iLumens Department, Paris Descartes University, Paris, France

Eppocrate, NGO, Paris, France
e-mail: cmonteil@eppocrate.org

© Springer Nature Switzerland AG 2019

A. André (ed.), *Digital Medicine*, Health Informatics,
https://doi.org/10.1007/978-3-319-98216-8_4

Using the magic notebook, a group of people can be insured to always be in sync with their notes. They could coordinate ideas, plans, or contracts for example. The magic notebook could also be used to send messages between people, with transparency and traceability.

Indeed this is how the Bitcoin Blockchain works. The Blockchain acts like the magic notebook and each time a Bitcoin is transferred, a transaction is recorded describing how much money was sent between multiple people. The Blockchain, like the magic notebook, insures that every copy of the transaction ledger is perfectly in sync.

4.1.2 To Get a Bit More Technical

We can now translate the metaphor of the magic notebook to the Blockchain, an open-source technology that was first used in 2009 and for which interest has been growing ever since. We will separate the Blockchain in three modules:

- Blocks
- Transactions
- Consensus

4.1.2.1 Blocks

The magic notebook is the Blockchain, and the pages are the blocks, which contain transactions. Each block is cryptographically linked to the previous and the next block, creating a chain of blocks that cannot be modified once created. Any attempt to modify or switch them would break the chain.

4.1.2.2 Transactions

Transactions are the content of the notebook. They can be monetary (cryptocurrencies) or not (messages, digital contracts, etc.). Like on the magic notebook, they are added in a chronological order and time-stamped.

When transactions are monetary, the amount of the transactions is displayed publicly. When it is a digital asset, it appears encrypted, so no one can see the original message.

Transactions are pseudo-anonymous. People use public and private keys to send and receive transactions. We can compare this key system to an email address that would be public and displayed (without mentioning a name) with a private password to decode the message and/or receive money.

Transactions are very useful to store "proofs of data." We can do so by securely encrypting data in a single piece of coded text (a "hash") and storing that proof on the Blockchain where it will be time-stamped. The holder of the original data can then at any point in time prove the veracity of the information. This can be extended to storing many proofs of data, creating centralized audit trails from multiple parties.

However, it is important to note that the Blockchain is not designed to store massive information inside transactions, but rather to store extensive small proofs of data (or a small message, a small piece of code, etc.) and/or exchange currencies, tokens, etc. The original information of digital assets (the written content of a book or a contract, for example) will stay on their original database, and the proofs will be stored in a transaction on the Blockchain.

4.1.2.3 Consensus

The Blockchain has no central authority like a President or a CEO. It is controlled by the people. Decisions are made by consensus from the community (>50% must agree for any decision to be taken). Therefore, the more people are involved, the more trustworthy the network becomes (which represents, e.g., for the Bitcoin Blockchain, tens of thousands of computers). As a result, any attempts of modifications to validated transactions will be rejected, ensuring the trust of the system.

Some Blockchains are called "public" when participants have full access and rights to the Blockchain (read, write, consensus decisions), and some are called "permissoned," when participants have different rights (some can only read, for example). A "private" Blockchain means that only a restricted and selected amount of people can participate to a Blockchain.

Finally, the magic notebook metaphor of having numerous copies that are always in sync, held by different people, represents a distributed network. A Blockchain is always distributed among a network of computers and constantly synchronizes itself on all the copies. This system prevents hackers from stealing the data of a Blockchain, as they would have to steal every copy on every computer having a copy.

4.1.3 The Bitcoin Blockchain

As an example, the Bitcoin Blockchain has its transactions system based on a cryptocurrency: Bitcoin. It is public and decentralized. The Bitcoin Blockchain was the first Blockchain protocol, but as it is an open-source protocol, there are now many more Blockchains, each one a bit different than the other!

4.2 The Uses of Blockchain

A Blockchain allows participants to share digital assets and information in a very secure way and without any third party involved. Every point of data entered is time-stamped, traceable, and auditable. In short, the main idea behind the Blockchain is to create a new form of trust when there is a need to trace and share data (information, processes, assets, currencies, etc.) across multiple parties with a high standard of security, with a possibility to be able to verify the integrity of all the data.

4.2.1 How Can Blockchain Impact the Health-Care Field?

We live in an era where medicine is at its best. We saw technological and knowledge breakthroughs greatly enhance the care we deliver to patients. However, our health-care system still holds many friction points. We will examine here cases where Blockchain can be a leverage to address some of health care's biggest challenges.

4.2.2 Clinical Trials

There is a growing consensus that the clinical trial system is undergoing a crisis. The saying "Publish or Perish" illustrates a system where to get funding, researchers have to publish rapidly and continually, promoting quantity over quality, and to publish studies yielding positive results. Only then can researchers make a name for themselves, facilitating an easier access to funding, hence to conduct researches. Of course, in reality, clinical trials take time and don't often yield positive results. If anything, the fact that all of someone's hypotheses actually become true should actually raise suspicion. This pressure to publish has been demonstrated to be linked with scientific fraud and questionable ethics to hype up results and increase chances of being published. This effect is also observed for studies funded by the pharmaceutical industry for obvious conflicts of interest. As a consequence, 80% of clinical trials are not reproductible today, violating the essence of science which is to ensure the veracity of findings by showing consistency in reproducting them, until someone proves them wrong.

Although there is great work to be done in changing the culture of ultimately judging researchers by the quantity of research they publish or the perspective on negative results as being as informative as positive results, the Blockchain can be a useful technology to build transparency into the research workflow.

How? By enabling the creation of an audit trail of clinical trials. Researchers can register their study design in the first place, and then record their progress in the study, from gathering the patient's consent to their inclusion, the data collected (questionnaires, biological samples, etc.), intermediary and final results, adverse effects, and so on. Any attempt to make modifications after the information is recorded (study design changes, result falsification, addition of fictitious patients or ghost authors, etc.) would make it more difficult to hide.

Blockchain can enable a built-in trial verification and ensure data integrity and transparency. It is an opportunity for clinical trials to be more transparent and reproductible, and to promote quality over quantity or positiveness of results. A very good example is the work done between the French hospital AP-HP and the Blockchain technology provider, Stratumn, presented in the publication "Blockchain protocols in clinical trials: Transparency and traceability of consent," [1] where the researchers proposed and implemented a method for notarizing participant consent for clinical trials, using the Bitcoin Blockchain.

4.2.3 Health-Care Medical Records

Electronic medical records (EMR) meet the double challenge to preserve patient privacy and to share health-care data among providers and researchers when needed. Many authors claim that health-care data should be owned and controlled by patients, instead of being scattered in different health-care systems. Yue et al. proposed an app (called Healthcare Data Gateway (HGD)) architecture based on Blockchain to enable the patient to own, control, and share their own data easily and securely without violating privacy [2]. Similarly, the start-up Iryo is building an appropriate platform for keeping health records unified. Instead of all kinds of medical data from various providers stored in different formats in different databases, Iryo's solution aims to store data securely and allow patients to share their medical history anywhere in the world [3]. Other solutions are given by fast-growing companies such as Patientory (the UK), Coral Health (Canada) [4], or Guardtime, an Estonian company who signed a deal in 2017 with Estonia's e-health authority to secure the health records of over a million Estonians [5]. It is important to note that the data in itself is not stored directly on the Blockchain, it is stored in classic databases. The Blockchain is used to create an audit trail of all modifications made in all the different places, detailing the who, what, when and where the information associated with each entry comes from, insuring immutability. It is not a replacement for the EMR but rather one technology among others that when combined, will yield better options. Blockchain, together with existing databases, access control systems and cryptography enables us to build next generation information systems that can securely manage patient information across organizational boundaries (hospitals, governments, etc.) while insuring traceability and accountability of the information.

4.2.4 Drug Manufacturing

In the pharma supply chain, safe and timely delivery is crucial for many products (vaccines, chemotherapies, etc.). As defended by the society Chronicled [6], and the MediLedger Project [7], Blockchain could help facilitate logistics, minimize discrepancies, create potential cost savings from streamlined processes, improve product visibility, and add transparency as products travel through the supply chain.

Between 2007 and 2013, the counterfeit drug market has increased by 300%, particularly with increased internet sales of pharmaceuticals. According to the

World Health Organization, roughly 50% of drugs consumed in developing nations are counterfeit, leading, depending on the sources, to 100,000–800,000 of deaths per year across the globe.

From the shape, size, and color of the medicine, to the packaging made to look exactly the same as the real medicine, these counterfeit pills contain smaller or no amounts of the original active ingredients or an improper mix of ingredients. Unfortunately, tracking and tracing the supply chain of drugs (shipping logistics, cold-chain management, etc.) are tedious and complex. To this day, regulatory authorities' attempts have been a struggle.

Blockchain technology can bring part of the solution by creating an audit trail of the entire supply chain of drugs, bridging the gaps between suppliers, vendors, distributors, and other intermediaries between the pharmaceutical company and the end user. This system would allow the final user or any regulatory authority to verify the origin of any drug and each step of its supply chain.

4.2.5 Public Health

Public health is the science of preventing disease and promoting human health. Analyzing the health of a population and the threats to health (disease outbreaks, alcohol, tobacco risks, etc.) is done through surveillance of cases, health indicators, and scientific research. Public health practice requires multidisciplinary teams of workers and professionals, and the collection and analysis on big volumes of data.

The usefulness of Blockchain relies in allowing numerous entities to share critical information across borders and in a timely manner. This would help to avoid the complexity of data exchange between non-interoperable entities, often involving manual procedures and long delays. The 2014 Ebola outbreak demonstrated the weaknesses of our current systems. When a few medical evacuees (contaminated people) were permitted back in the USA because the US government did not get their health information in time, it led to a rapid rise of the virus in the country. A Blockchain-based system could participate to prevent such cases.

In the USA, the Center of Disease Control and Prevention has partnered with IBM to develop a Blockchain-based solution to improve data collection, analysis, and sharing at the federal level, involving health agencies and hospitals. The objective is to improve collaboration through secure, compliant, and transparent data exchange systems, with the expectation that the data from a disease outbreak reach the required individuals so as to control the spread of the infection as fast as possible.

The start-up SimplyVital Health, launched a Blockchain platform built as an audit trail, where care providers can view the same data for shared patients. Financial and clinical algorithms powered by artificial intelligence (AI) provide actionable strategic opportunities for all users. As its CEO says "Together, providers work to drive down the cost of care, which means providers are paid for working together and for results. Because reimbursement depends on the ability to prove they worked together, the immutable audit trail provides extra upside for the users." [8].

In the future genomics market, the guarantee of privacy and control would be essential. Nebula Genomics argues that this market will have similar characteristics and challenges as any market where data is involved. Blockchain might be the appropriate technology to solve data integrity issues and ensure data can be validated without any third party participant [4].

4.3 Conclusion

While the technology holds great potential to increase efficiency in many areas of healthcare, it isn't an end-to-end solution. Combined with other technologies, it will be a very useful tool for many compelling use cases, such as those discussed here and many more. We are still in the early days of discovering how we can best harness this promising technology and healthcare systems are notorious for being tedious to revolutionize but the quest is definitely worth it for patients.

References

1. Benchoufi M, Porcher R, Ravaud P. Blockchain protocols in clinical trials: Transparency and traceability of consent [version 1; referees: 1 approved, 1 not approved]. F1000Research. 2017;6:66. https://doi.org/10.12688/f1000research.10531.1.
2. Yue X, Wang H, Jin D, et al. Healthcare data gateways: found healthcare intelligence on blockchain with novel privacy risk control. J Med Syst. 2016;40(10):218. https://doi.org/10.1007/s10916-016-0574-6. [published Online First: 2016/08/26].
3. https://iryo.io/.
4. @berci. Top 12 companies bringing blockchain to healthcare – The Medical Futurist. 2018.
5. Estonia prescribes blockchain for healthcare data security – Health Matters. 2018.
6. Paid program: block chain meets cold chain. 2017.
7. MediLedger – blockchain solutions for pharma companies. 2018.
8. Blockchain in healthcare: 2017's successes – CoinDesk. 2017. https://www.coindesk.com/blockchain-healthcare-2017s-successes/.

Chapter 5
Precision Medicine

Arthur André and Jean-Jacques Vignaux

5.1 From Reductionism to a Systems Approach

Modelizing biology and life has always been a challenge to the modern scientific method due to the complexity of the components and interactions of a biological organization from the cell to a whole human body. The reductionist approach of biological organization, with the help of traditional mathematics and physics, has successfully identified many of the components and many of the interactions but, unfortunately, offers no convincing concepts or methods to understand how system properties emerge. Information technology (IT) offers the ability to treat and organize large amounts of data and leads to a paradigm of integration—in opposition to reduction—to explain biological systems and phenomenon [1]. Quantitative datasets of DNA, RNA, proteins, and metabolites provide an unprecedented starting point to understand the effects of perturbations on a cell [2] and, with addition of clinical tests and imaging, the effect on the whole body. The informational view of biology defines biological information—biomarker—as a given data integrated in a network. This leads to a "systems" approach to physiology and pathophysiology. Some of the pioneers were undoubtedly Nobel prize winners Alan Lloyd Hodgkin and Andrew Fielding Huxley, who constructed the first numerical simulation that explained the action potential propagating along the axon of a neuronal cell in 1952 [3]. Eight years later, Denis Noble applied a similar method to heart cells and gave his definition to this emerging domain: "Systems biology ... is about putting together rather than taking apart, integration rather than reduction. It requires that we develop ways

A. André (✉)
Pitié-Salpêtrière Hospital, Assistance Publique Hôpitaux de Paris, Paris, France

Sorbonne Université, Paris, France
e-mail: arthur.andre@neurochirurgie.fr

J.-J. Vignaux
iTechCare Medical Data Research, Osteopath, Paris, France

© Springer Nature Switzerland AG 2019

A. André (ed.), *Digital Medicine*, Health Informatics,
https://doi.org/10.1007/978-3-319-98216-8_5

of thinking about integration that are as rigorous as our reductionist programmes, but different. ... It means changing our philosophy, in the full sense of the term" [4].

5.2 Systems Biology and Systems Medicine

Systems biology is then a holistic approach to deciphering the complexity of biological systems that starts from the understanding that the networks, which form the whole of living organisms, are more than the sum of their parts (so-called reductionism). It is collaborative, integrating many scientific disciplines—biology, computer science, engineering, bioinformatics, physics, and others—to predict how these systems change over time and under varying conditions [5].

The formal study of systems biology, as a distinct discipline, was launched by system theorist Mihajlo Mesarovic in 1966 in an international symposium at the Case Institute of Technology in Cleveland, Ohio, titled "Systems Theory and Biology" [1, 6]. The concept has been since used widely in biology in a variety of contexts. The Human Genome Project is an example of applied systems thinking in biology which has led to new, collaborative ways of working on problems in the biological field of genetics [7, 8].

The earliest uses of the term *systems medicine* appeared in 1992, in an article on the *holographic model of human body* by B.J. Zeng [9] and in a paper on systems biomedicine by T. Kamada [10]. Indeed, systems medicine is another multidisciplinary field of study that looks at the systems of the human body as part of an integrated whole, incorporating biochemical, physiological, and environment interactions. This ability to design predictive, multiscale models enables scientists to discover new biomarkers for disease, stratify patients based on unique genetic profiles, and target drugs and other treatments [5].

The transformative power of systems biology was institutionally recognized in a 2009 report from the National Research Council titled "A New Biology for the 21st Century: Ensuring the United States Leads the Coming Biology Revolution." The Committee reported that years of research "have generated detailed information about the components of the complex systems that characterize life – genes, cells, organisms and ecosystems – and this knowledge has begun to fuse into greater understanding of how all those components work together as systems" [11, 12].

With the present emerging artificial intelligence (AI) technologies, computational "machine learning" techniques (*see* Chap. 3) for training and generalization from data, and cutting-edge statistical techniques, will play a significant role in analyzing multidimensional datasets generated by the new technologies of systems medicine [13, 14]. This progressively but rapidly leads to a "new taxonomy," generating new approaches for disease diagnosis, therapy, and clinical decisions, promising more individualized treatments and improved outcomes for patients [12]. Indeed if this approach becomes efficient in clinical practice, it provides a real paradigm change in health care, from reactive to proactive medicine. This proactive approach is already very promising in "long-term" management diseases: chronic illnesses [15], infectious diseases [16], or cancer [17, 18].

5.3 "4P" Medicine

The concept of "4P" medicine, that is predictive, preventive, personalized, and participatory, has been advocated by Leroy Hood and colleagues since more than a decade now.

Actually, the P4 medicine concept has emerged from the convergence of three trends [14]:

- The increasing ability of systems biology and systems medicine to decipher the biological complexity of disease: Each pathology is seen as a personal experience and a dynamic network which is dysregulated from health to disease, and conversely treatment is a pathway from disease to health, considering each personal dimension from molecular to social and its participation in the pathological state.
- The increasing ability of computers to integrate, store, analyze, and communicate data from medical records, symptoms, clinical tests, biological samples, imaging, and molecular biology: Then, a personalized data cloud can be imagined, including multiple dimensions of each individual, from genetic to phenotypic characteristics, but also sociometrics (social level, education, familial context, etc.).
- The increasing access to information for patients and consequently their interest in managing their own health: The digital tools not only connect recent science fundamental discoveries to clinical applications but also the patients and healthcare consumers. In 2013, one in three Americans had gone online to investigate a medical condition [19]. This phenomenon is probably increasing.

Drug research could also benefit from these tools. Indeed, the omics data analysis would not only allow the identification of new biomarkers representative of pathologies but also propose new therapeutic targets for the development of more effective drugs. Thus, thanks to the analysis of 70,000 scientific articles by an AI [20], proteins associated with numerous cancers could be identified. This discovery would have taken years if it had been done by human means, as far as it could be made.

5.4 How to Build a Systems Medicine?

5.4.1 Symptoms at the Heart of Integrative Analysis

Symptoms are disorders expressed and felt by patients and are at the very heart of disease recognition. They are the clinical reflections of the alteration of physiological functions. When different physiological functions degenerate, symptoms can coevolve together. This is why the reading of clinical comorbidity (Fig. 5.1: Comorbidity patterns in patients with chronic diseases in general practice ©) through the association with symptoms is crucial to interpret a pathological state: we call it *network medicine* [21].

EDC	No. Cases	ID only	ID+1 or more	ID+2 or more	ID+3 or more	ID+4 or more	ID+5 or more
PATTERN A							
Hypertension (with and without complications) [CAR1415]	28760	11.05	88.95	67.95	44.05	24.45	12.10
Disorders of lipid metabolism [CAR11]	22345	13.19	86.81	65.76	43.29	24.58	12.49
Type 2 diabetes (with and without complications) [END0607]	10058	9.67	90.33	74.55	54.11	33.86	18.86
Cardiac arrhythmia [CAR09]	5777	8.60	91.40	78.43	61.28	42.08	25.36
PATTERN B							
Cerebrovascular disease [NUR05]	2658	5.91	94.09	83.07	66.59	46.46	28.33
Ischemic heart disease (excluding AMI) [CAR03]	2344	3.97	96.03	86.90	70.56	49.40	32.81
Chronic renal failure [REN01]	1964	5.86	94.14	84.57	69.25	51.12	33.50
Congestive heart failure [CAR05]	1377	3.05	96.95	90.12	78.29	61.26	42.19
PATTERN C							
Anxiety and depression [PSY0109]	27357	36.46	63.54	39.54	24.38	13.91	7.14
Thyroid disease [END04]	19299	31.97	68.03	42.49	26.26	15.12	7.86
Asthma [ALL04]	7614	40.74	59.26	34.98	21.55	13.12	7.17
Schizophrenia and affective psychoses [PSY07]	1309	31.02	68.98	44.16	24.68	13.67	7.87
PATTERN D							
Obesity [NUT03]	19640	17.11	82.89	60.58	39.76	22.68	11.36
Osteoporosis [END02]	6143	9.56	90.44	72.54	49.44	29.01	14.76
Deafness, hearing loss [EAR08]	5403	18.88	81.12	61.08	42.88	27.63	15.57
Malignant neoplasms [NEOMAL]	5138	14.25	85.75	66.62	46.44	28.67	15.84
Degenerative joint disease [MUS03]	4452	11.25	88.75	72.24	52.34	32.79	17.65
Bengin prostatic hypertrophy [GUR04]	4089	11.49	88.51	68.38	45.98	27.42	14.97
Emphysema, chronic bronchitis, COPD [RES04]	3183	9.80	90.20	72.86	54.16	35.38	21.08
Generalized atherosclerosis [CAR10]	2705	13.20	86.80	72.16	55.75	39.26	23.51
Glaucoma [EYE08]	2450	7.31	92.69	79.47	60.98	32.81	23.80
Chronic liver disease [GAS05]	2121	13.11	86.89	67.61	48.33	33.81	18.39
Dementia and delusions [NUR11]	1112	11.33	88.67	73.38	52.25	33.45	19.06
Chronic skin ulcer [REC03]	955	8.90	91.10	73.49	58.95	41.05	24.91
Cardiac valve disease [CAR06]	936	7.69	92.31	82.16	64.10	47.54	30.02
Parkinson's disease [NUR06]	805	9.57	90.43	77.14	55.53	38.63	26.09

EDC: expanded diagnosis cluster; ID: index disease; COPD: chronic obstructive pulmonary disease; AMI: acute myocardial infarction.
doi:10.1371/journal.pone.0032141.t001

Fig. 5.1 Comorbidity patterns in patients with chronic diseases in general practice ©

The analysis of comorbidities (diseases that accompany others) allows the emergence of groups of data or labels for diagnosis [22]. These datasets help identify dynamic interactions at different scales between different medical data [23].

5.4.2 Symptom Clusters

The grouping of symptoms in a cluster remains controversial. Indeed, without the computation capacity of computers, clustering (grouping family data) only makes sense if the cluster is closed, for example, the precise clinical description of a disease. However, thanks to computational methods, we can de-compartmentalize clusters to integrate more distant data. Since the early days of medicine, closed cluster use has highlighted the relationship between signs and symptoms within a single clinical syndrome. They allow, in the clinical setting, to identify a disease by comparing the symptoms evoked with those of the clinical syndrome, which defines the disease theoretically.

Today the stakes are different. The population is aging and life expectancy is increasing, chronic diseases are developing, and several diseases can coexist in the same patient. It is therefore necessary to de-compartmentalize groups of data associated with one disease in favor of a broader, integrative medicine, allowing several diseases to be included in new clusters.

Data coevolution makes it possible to envisage and anticipate comorbidities. The important comorbidity between cardiovascular, metabolic, musculoskeletal, and digestive pathologies, for example, requires an exploration that lifts the barriers of clinical disciplinary pictures in order to forecast the evolution of health in a more general sense.

Marylin J. Dodd [24] draws a portrait of the clustering of symptoms by gradually returning to the history of the method and its issues today thanks to current technological skills. This article leads to a reflection on the method of management of symptomatological data and especially the standardization of the integration pathway and interoperability [25]. Interoperability allows an algorithm architecture to be integrated so that, on the one hand, the notion of clustering is preserved and, on the other hand, to ensure that the data are linked to each other when new data is integrated.

5.4.3 Comorbidity

The analysis of comorbidity is one of the major challenges of bioinformatics today. Computing powers, conceptualization of biological systems, big data, connected health, and telemedicine are all arguments for interpreting a broader sense of health. This "whole-body system" understanding would make it possible to provide a personalized preventive analysis.

A non-randomized experiment has presented a clustering of symptoms that can refine the understanding of acute and chronic pain [26]. This study illustrates the benefit that could be obtained by interpreting all the symptoms independently of a specific pathological context.

More recently, a study carried out on a large cohort contains the inventory of comorbidities in the daily clinical practice [27]. Figure 5.1 supports the idea of an organism that evolves in parallel in each system (Fig. 5.1. Comorbidity patterns in patients with chronic diseases in general practice ©). Indeed, four patterns of diseases developing jointly are defined. Each pattern is associated with a risk of having a comorbidity. For example, in PATTERN B, in cerebrovascular disease, risk of developing ischemic heart disease, kidney failure, or cardiac hemorrhagic diseases is important. We talk about the high prevalence of comorbidity, unlike PATTERN C where there is a low comorbidity burden.

One can understand that it is relevant to apprehend pathological burdens jointly rather than to be restricted to a single disease analyzed in isolation from the others in the same patient.

This association approach has to be achieved, not only at the level of symptoms but also by integrating other data that may have their interest to ensure the validity of a diagnosis and a therapeutic approach: biological, genetic, connected health, etc. datasets.

5.4.4 *Complex Systems in Integrative Medicine*

Complex systems in the field of biology, a cluster of symptoms in the field of clinical diagnosis, grouped in "network medicine," can take many forms.

They are described at the fundamentals of bioinformatics and are presented as a family of biomedical data coevolving together. Let's take the example of biomarkers of Alzheimer's disease. We know that they are associated with each other by their affiliation to the same disease. This complex system makes it possible to develop algorithms for the recognition, as early as possible, of physiopathological tendencies in Alzheimer's disease, as well as cardiovascular diseases, autism, diabetes, obesity, certain cancers, but also in the pathophysiology of aging [28].

These diseases are complex because of the heterogeneity and the great variability of the probable etiologies, often co-factorial, notably because of the impact of epigenetic mechanisms and environmental factors [29].

Another example concerns autism. Diaz-Beltran et al. explored autism using an interdisciplinary biological approach and highlighted the genetic, environmental, and immunological combination added to the neurological factors that may be at the origin of the autistic spectrum. They have developed an integrative approach that is useful for his diagnosis [30].

5.5 Biomarkers

A biomarker is an objectively measurable characteristic that represents an indicator of normal and pathological biological processes or an indicator of pharmacological response following a therapeutic intervention [31].

The identification of biomarkers linked to pathologies makes it possible to collect a large number of data that are useful for determining chronobiological trends. It is by submitting this data to a dynamic analysis in bioinformatics that the medicine will be able to propose diagnosis and prognosis of great precision which is always more personalized.

Wehling et al. describe the discovery and analysis of biomarkers as the most appropriate field of research to assist clinical medicine [32]. Biomarkers can collect measurements of different physiological activities for the benefit of their interpretations under pathological conditions [33].

Many factors can affect biomarker values. To ensure the sensitivity and specificity of biomarker evaluation, systems biology methods integrating translational bioinformatics and data processing strategies have been developed. The goal of identifying new biomarkers is to support the medical decision. Indeed, it is through the identification of more accurate and robust biomarkers or the identification of groups of biomarkers that tomorrow's physicians will be better able to study the many multidimensional factors of health dynamics.

5.6 Bioinformatics

The needs of clinicians have conditioned the required specificities of new methods of bioinformatics. Databases have to be:

- Dynamic: the database must be automated and scalable as new clinical cases are experienced, as well as advances in applied and clinical research.
- Interactive: it must be used at the heart of the consultation with an easy-to-use interface.
- From global to specific: the new integrated data must be quickly usable by the clinician regardless of the origin of the data (for example when integrating data from m-health devices).
- Integrating standardized data with respect to the medical objective, to ensure the interoperability of a given data with another dataset.
- Combining data with each other by the automated artificial intelligence experiment: to quickly process the addition of data and make them clinically relevant by associating them with an already structured database, it is essential to have an algorithm that is sufficiently integrated.

5.6.1 Current Integration Tools

The consolidation of complex adaptation systems is one of the main challenges of bioinformatics research. In order to build theoretical systems in biology and guide research in bioinformatics, it is necessary to have thorough knowledge of principles in genomics, proteomics, molecular biology, clinical science, and phenomenal systems.

Many software programs today rely on databases. Nowadays any new database has his own associated software to address integration and standardization issues (see Table 6.1 Biomarkers in Precision Medicine: The Era of Omics). Today, they are mostly used in applied research and especially in the very wide field of omics. Some technological innovations are demonstrating their success in accompanying patients in heavy care episodes, especially in surgery or oncology [34]. More recently, academic initiative are exploring translational medicine [35, 36]. The objective of this transversal medicine is to transform the results of basic research in life science into new tools and methods in clinical management. Today the great variability and diversity of exploitable medical data make it difficult to standardize and then integrate, notably since it is necessary to bring together all the medical and biomedical experts to promote their interpretation. It is thus necessary to consolidate suitable integration methods and paths. In most projects initiated, technologies are arranged in series to ensure the exploitation of data. Indeed, a specialized web host, operating system and databases are associated to provide medical experts with an overview of the care episode. The National Institutes of Health (NIH) has benefited from all the new technologies that can be used to launch a program called "Big Data to Knowledge"

(BD2K) [37]. In fact, the NIH's ambitionis to exploit large-scale health data by implementing an IT initiative that would consolidate training programs tailored to new medical experts. The NIH BD2K program seeks to position health data science at the heart of biomedical research in the 21st century. Other areas exist in the data treatment concerning rare diseases. In Europe, Orphanet and Inria have collaborated to create Orphamine, a tool for the collection and processing of rare diseases data, allowing general visualization of Orphanet data for specialists [38]. Finally, the Internet giants are also at the forefront of health innovation, especially Google, which stands out in data science. Indeed, they seek to exploit health data to know the daily health conditions of patients, and to move from rare and monogenic models to common diseases. It is notably Google's verily, Duke an Stanford Universities who are involved in such research [39]. However, those pathologies are much more multifactorial and polygenic, and this is nowadays still a complex challenge [40].

5.7 Limits

Despite the great promises of big data and omics interpretation, many experts agree to admit that these computational tools will not be sufficient to meet the challenge of deciphering biological complexity. Domain expertise in biology is essential, where decades of research can be leveraged to help interpret this data [41]. As Kohler et al. says, "Without a deep and growing understanding of biological phenomena and networks, it will not be possible to find the critical signals in the tremendous noise generated by vast heterogeneous data" [41]. Moreover this ubiquitous definition of self for every human being could lead to an "overmedicalization" effect, with potential harms of excessive preventive measure [42].

References

1. Systems biology. Wikipedia. 2018.
2. Sauer U, Heinemann M, Zamboni N. Getting closer to the whole picture. Science. 2007;316:550. https://doi.org/10.1126/science.1142502.
3. Hodgkin AL, Huxley AF. A quantitative description of membrane current and its application to conduction and excitation in nerve. J Physiol. 1952;117(4):500–44.
4. Noble D. Cardiac action and pacemaker potentials based on the Hodgkin-Huxley equations. Nature. 1960;188:495–7. https://doi.org/10.1038/188495b0.
5. @ISBUSA. What is systems biology. Institute for Systems Biology; Seattle, 2018.
6. Rosen R. Systems theory and biology. In: Mesarović MD, editor. Proceedings of the 3rd systems symposium, Cleveland, Oct 1966. Springer, New York; 1968. xii + 403 p., illus. $16. 1968. https://doi.org/10.1126/science.161.3836.34
7. Zewail A. Physical biology: from atoms to medicine. London: Imperial College Press. p. 339.
8. Human Genome Project. Wikipedia. 2018.
9. Zeng BJ. On the holographic model of human body. In: 1st national conference of comparative studies traditional Chinese medicine and west medicine, medicine and philosophy, 1992.

10. Kamada T. System biomedicine: a new paradigm in bio-medical engineering. Jpn J Med Electron Biol Eng. 1991;29(Supplement):1–1.
11. National Research Council Committee on a New Biology for the 21st Century: Ensuring the United States Leads the Coming Biology R. The National Academies Collection: Reports funded by National Institutes of Health. A new biology for the 21st century: ensuring the United States leads the coming biology revolution. Washington, DC: National Academies Press (US)National Academy of Sciences; 2009.
12. National Research Council Committee on AFfDaNToD. The National Academies Collection: Reports funded by National Institutes of Health. Toward precision medicine: building a knowledge network for biomedical research and a new taxonomy of disease. Washington, DC: National Academies Press (US)National Academy of Sciences; 2011.
13. Schadt EE, Linderman MD, Sorenson J, et al. Computational solutions to large-scale data management and analysis. Nat Rev Genet. 2010;11(9):647–57. https://doi.org/10.1038/nrg2857. [published Online First: 2010/08/19].
14. Flores M, Glusman G, Brogaard K, et al. P4 medicine: how systems medicine will transform the healthcare sector and society. Per Med. 2013;10(6):565–76. https://doi.org/10.2217/pme.13.57.
15. Sagner M, McNeil A, Puska P, et al. The P4 health spectrum – a predictive, preventive, personalized and participatory continuum for promoting healthspan. Prog Cardiovasc Dis. 2017;59(5):506–21. https://doi.org/10.1016/j.pcad.2016.08.002. [published Online First: 2016/08/23].
16. Bengoechea JA. Infection systems biology: from reactive to proactive (P4) medicine. Int Microbiol. 2012;15(2):55–60. https://doi.org/10.2436/20.1501.01.158. [published Online First: 2012/08/01].
17. Tian Q, Price ND, Hood L. Systems cancer medicine: towards realization of predictive, preventive, personalized and participatory (P4) medicine. J Intern Med. 2012;271(2):111–21. https://doi.org/10.1111/j.1365-2796.2011.02498.x. [published Online First: 2011/12/07].
18. Lu M, Zhan X. The crucial role of multiomic approach in cancer research and clinically relevant outcomes. EPMA J. 2018;9(1):77–102. https://doi.org/10.1007/s13167-018-0128-8. [published Online First: 2018/03/09].
19. Fox S, Duggan M. Health online 2013. Washington, DC: Pew Internet & American Life Project; 2013. p. 1.
20. Picton G. Study shows promise in automated reasoning, hypothesis generation over complete medical literature 2018. Available from: https://www.bcm.edu/news/research/automated-reasoning-hypothesis-generation.
21. Watts D, Newman M, Barabási, A. The structure and dynamics of networks Princeton studies in complexity Mathematics -Applied (Paperback and eBook). 2006.
22. Yan Q. Translational bioinformatics and systems biology methods for personalized medicine. 1st ed: PharmTao, Santa Clara, CA, USA; 2018.
23. Iris F. Biological modeling in the discovery and validation of cognitive dysfunctions biomarkers. In: Turck C, editor. Biomarkers for psychiatric disorders. Boston: Springer; 2008.
24. Dodd MJ, Miaskowski C, Lee KA. Occurrence of symptom clusters. JNCI Monogr. 2018;2004(32):76–8. https://doi.org/10.1093/jncimonographs/lgh008.
25. R. C. Partage de données biomédicales : modèles, sémantique et qualité. BioInformatique. Biologie Systémique. Université Pierre et Marie Curie – Paris VI: Université Pierre et Marie Curie – Paris VI.
26. Fishbain D, Gao JR, Lewis J, et al. Examination of symptom clusters in acute and chronic pain patients. Pain Physician. 2014;17(3):E349–57. [published Online First: 2014/05/23].
27. Garcia-Olmos L, Salvador CH, Alberquilla A, et al. Comorbidity patterns in patients with chronic diseases in general practice. PLoS One. 2012;7(2):e32141. https://doi.org/10.1371/journal.pone.0032141. [published Online First: 2012/02/24].
28. Cho DY, Kim YA, Przytycka TM. Chapter 5: Network biology approach to complex diseases. PLoS Comput Biol. 2012;8(12):e1002820. https://doi.org/10.1371/journal.pcbi.1002820. [published Online First: 2013/01/10].

29. Suderman M, Hallett M. Tools for visually exploring biological networks. Bioinformatics. 2007;23(20):2651–9. https://doi.org/10.1093/bioinformatics/btm401. [published Online First: 2007/08/28].
30. Diaz-Beltran L, Cano C, Wall DP, et al. Systems biology as a comparative approach to understand complex gene expression in neurological diseases. Behav Sci (Basel). 2013;3:253–72.
31. Strimbu K, Tavel JA. What are biomarkers? Curr Opin HIV AIDS. 2010;5(6):463–6. https://doi.org/10.1097/COH.0b013e32833ed177. [published Online First: 2010/10/28].
32. Wehling M. Translational medicine: science or wishful thinking? J Transl Med England. 2008;6:31.
33. Yan Q. Toward the integration of personalized and systems medicine: challenges, opportunities and approaches. 2010 Pers Med. 2011;8(1):1–4. https://doi.org/10.2217/pme.10.77.
34. Collins FS, Varmus H. A new initiative on precision medicine. N Engl J Med. 2015;372(9):793–5.
35. Durako AR, Mishkel G. Creating the virtual cardiac surgical home: implementation of an app based enhanced recovery after surgery platform to improve outcomes and patient engagement. J Am Coll Cardiol. 2018;71(11):A2121.
36. Satagopam V, Gu W, Eifes S, et al. Integration and visualization of translational medicine data for better understanding of human diseases. Big Data. 2016;4(2):97–108. https://doi.org/10.1089/big.2015.0057. [published Online First: 2016/07/22].
37. Bui AAT, Van Horn JD. Envisioning the future of 'big data' biomedicine. J Biomed Inform. 2017;69:115–7. https://doi.org/10.1016/j.jbi.2017.03.017.
38. Inria. OrphaMine – Plateforme de fouille de données pour les maladies rares – Inria. 2018.
39. Manos D. Rocking the baseline: Verily, Duke, and stanford aim to make medicine more predictive with a new baseline study. Clinical OMICs. 2017;4(3):3–4.
40. Wray NR, et al. Common disease is more complex than implied by the core gene omnigenic model. Cell. 2018;173(7):1573–80.
41. Philippi S, Köhler J. Addressing the problems with life-science databases for traditional uses and systems biology. Nat Rev Genet. 2006;7(6):482. https://doi.org/10.1038/nrg1872.
42. Vogt H, Hofmann B, Getz L. The new holism: P4 systems medicine and the medicalization of health and life itself. Med Health Care Philos. 2016;19(2):307–23. https://doi.org/10.1007/s11019-016-9683-8. [published Online First: 2016/01/29].

Chapter 6
Biomarkers in Precision Medicine: The Era of *Omics*

Jean-Jacques Vignaux and Arthur André

Since the advent of genomics, the suffix "-omics" has been added to the names of many fields to refer to large-scale studies to identify biomarkers. The -omics data analysis would not only allow the identification of new biomarkers representative of pathologies but also propose new therapeutic targets for the development of more effective drugs.

Identifying effective biomarkers for prognosis and diagnosis has been considered as one of the greatest challenges in recent years. It allows for a cross-disciplinary medicine between biomedical knowledge and the clinic. However, the complexity of the diseases and their heterogeneity make the discovery and validation of efficient biomarkers very difficult. To make the interpretation of biomarkers useful and applicable in a clinical context, it is necessary to ensure the reliability of the methods employed from their detection until their diffusion. The identification of new biomarkers goes through different stages [1]:

- Discovery
- Qualification
- Verification
- Optimization of clinical trials
- Clinical validations
- Diffusion

J.-J. Vignaux
iTechCare Medical Data Research, Osteopath, Paris, France

A. André (✉)
Pitié-Salpêtrière Hospital, Assistance Publique Hôpitaux de Paris, Paris, France

Sorbonne Université, Paris, France
e-mail: arthur.andre@neurochirurgie.fr

© Springer Nature Switzerland AG 2019

A. André (ed.), *Digital Medicine*, Health Informatics,
https://doi.org/10.1007/978-3-319-98216-8_6

6.1 Genomics

The computer methods developed in genetic research have made it possible to build a large database trying to gather and map all the discoveries in the field of DNA–protein interactions. Broad-spectrum chromatin immunoprecipitation (ChIP) [2] methods associated with high-throughput sequencing (HTP) technologies have allowed for rapid DNA sequencing. Genotype–phenotype correlations have thus been identified and currently constitute databases frequently used in genomic research or any other type of research known as "omics" (since the advent of genomics, the suffix "-omics" has been added to many field names for large-scale or genome-wide studies).

All the data and results obtained can be consulted on various databases dedicated to Integrative Clinical Medicine (Table 6.1: Computer databases).

Table 6.1 Computer databases

Datasets	Contents
Genomics	
ArrayExpress	Genomics sequences
BLAST	Similar genomics sequences
Gene Expression Omnibus	Public functional for genomics database
CLUSTAL	Genomics sequences
The 1000 Genomes Project	Public functional for genomics database
exPASy	Provide access to scientific databases and software tools
Metabolomics	
The Human Metabolome DataBase	Small molecules metabolites database
IntAct	Molecular interaction database
isoMETLIN	Small molecules metabolites database
Cytoscape	Molecular interaction network and biological pathways
Phenomics	
dbSNP	Public domain archive for genetics polymorphisms
The Human Variome Project	Human genetic variation affecting human health database
Human Phenotype Ontology	Public functional for genotype-phenotype association database
Ensembl	Variant effect predictor
Proteomics	
UniProt (2004*)*	Protein sequences and functional information (database integrating Swiss-Prot /TrEMBL/PIR-PSD)
Swiss-Prot	Protein sequences (540,000 annotated and reviewed)
TrEMBL	Protein sequences (50 million annotated and unreviewed)
RefSeq	DNA/RNA/protein sequences (virus, archaea, bacteria, and eukaryotes)
Cancer Proteomics DataBase	Proteomics
Human Protein Atlas	Spatial distribution of the human proteome
The Protein Data Bank (PDB)	Tools about 3D structure proteins, nucleic acids, and metabolites

Table 6.1 (continued)

Datasets	Contents
microRNA database	
miRBase [27]	miRNA sequences (avec annotations)
miRGate [28]	Database of human, mouse, and rat miRNAs/mRNAs targets
miRTarBase [29]	miRNA-target interactions (validated by an experimental method especially in micro-network)
miRDB [30]	miRNA-target predictions (several species)

In the context of rare diseases, the OMIM (Online Mendelian Inheritance in Man) database, which lists the associated human genes and phenotypes, has more than 8000 phenotypes, of which about 6000 molecular bases responsible for hereditary disorders have been identified and close to 4000 for which the gene involved is known [3].

A genome mapping work to identify the function of a gene in relation to its positioning on the 23 human chromosomes has taken place during the last three decades. This mapping is based on probabilities of a gene's impact on the phenotype, predominantly or recessively, and its position relative to other genes. Indeed, it is commonly accepted that a gene is closely related to the associated nucleotide sequences. Among the first techniques used, the method called "restriction fragment length polymorphism" (RFLP) has transformed the approach of genetic mapping. This technique using gel electrophoresis and a DNA probe initiated a race toward the identification of genes responsible for diseases such as Huntington's disease [4] or Duchenne muscular dystrophy [5].

The first reference mapping was published in 1987 [6], with RFLP markers for more than ten million nucleotides. This map represents the beginning of the databases. Then came more complex methods, based on physical mapping and the search for clones where the DNA was then fragmented and cloned, and then classical sequencing methods that allowed biologists to read short fragments and then link them together by bioinformatics to obtain a reconstitution of the DNA molecule.

Today, all the studies are gathered around the constitution of a back-up database providing a general vision of the human phenotype. This is the "GWAS," or genome-wide association studies, enabled by advanced techniques in genetic sequencing and computational methods, one of whose goals is to rapidly detect phenotypes. Today, more than 1000 GWAS have been published to identify more than 4000 genetic loci in relation to 500 phenotypic or pathological traits [7] (Fig. 6.1: GWAS method).

Moreover, studies conducted by Hebbring et al. [8] have drawn the interest of a phenotype-based approach. The latter makes it possible to associate different genetic mutations at the origin of a single phenotypic trait, by relying on current health data (signs, symptoms, or phenotypic manifestations) found in sets of EHR (electronic health record). The latter thus developed the "Phenome-wide association studies" (PheWAS) as a genotype–phenotype approach. This method made it possible to carry out a haplotype mapping (group of alleles of different loci located on

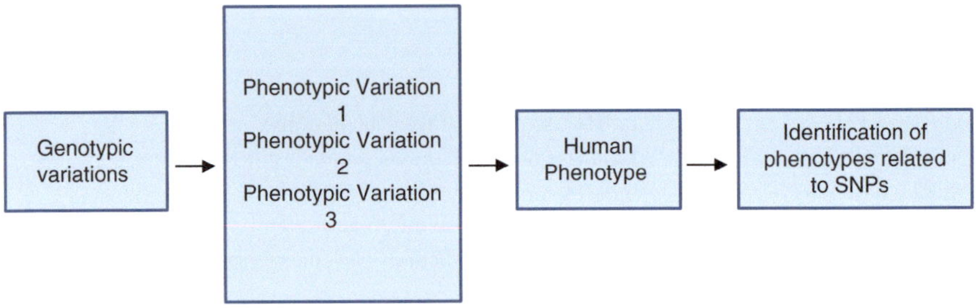

Fig. 6.1 GWAS method

the same chromosome and usually transmitted together) of the human genome called "HapMap."

The GWAS approach allowed, through the analysis of all single nucleotide polymorphisms (SNPs) in association with identified diseases, to centralize more than 38 million polymorphic variants in a database of 100,000 nuclear genes.

These databases are particularly relevant in the analysis of phenotypic manifestations or to establish the prevalence of pathologies.

Open-source databases (publicly available computer licenses) have been developed to promote exchange and collaborative work, such as SNOMED CT [9], Ontobee [10] or Human Phenotype Ontology [11], and the Diseases Database [12], all of which meet the Health Insurance Portability and Accountability Act (HIPAA) health data processing and exchange criteria approved by the United States Congress in 1996 for health and health insurance [13]. These databases also meet the technical criteria for programming databases of the International Classification of Diseases: 10th revision (ICD-10) that facilitate diagnosis.

Establishing a technical standard for databases promotes international collaboration. However, the need for completeness of genomic databases, during integration, can decentralize its use in practical cases specialized in the daily clinic. Tools of this scale are solid databases that facilitate biomedical research but must, in part, be restructured and oriented to serve the clinical management of the physician.

We can take the example of the *Human Phenotype Ontology* database whose role is to contribute to the diagnosis. This database makes it possible to associate genetic data with phenotypic elements with each other by means of lexical recognition (Fig. 6.2: HPO lexical recognition tools).

6.2 Proteomics

Protein is the product of gene expression: by its activities, it represents the functional aspect and the dynamics of the cell. Studying the variations in the amount of a protein from its synthesis to its function makes it possible to deduce cellular homeostasis with great precision. The proteomic level,

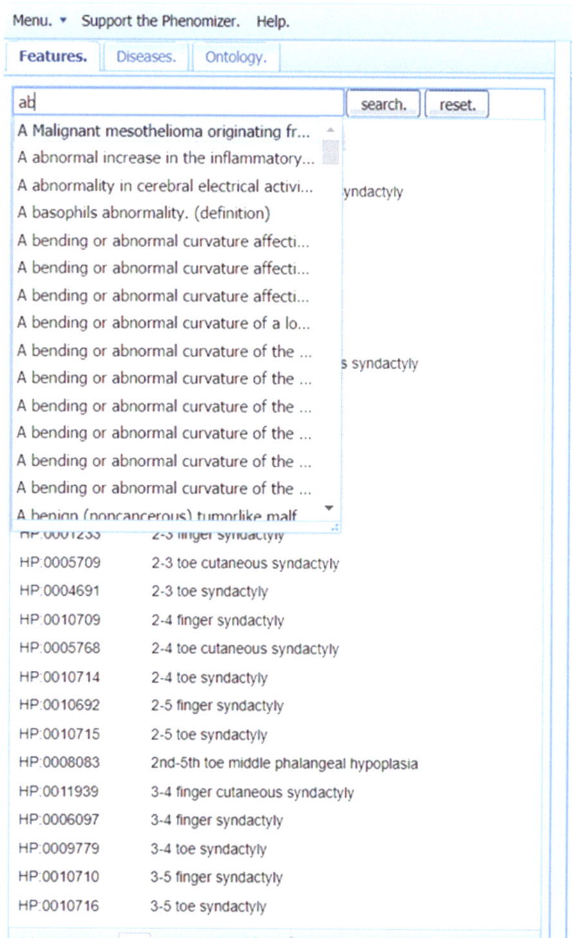

Fig. 6.2 HPO lexical recognition tools

combined with the metabolomic level, is a very dynamic parameter. It is therefore interesting to study protein biomarkers representative of various cellular activities [14] and their dysfunction, for example, neuromuscular degeneration in Duchenne's disease [15].

The screening and management of cancers also benefit from proteomic biomarker research, particularly by the early administration of drug adjuvants in locally advanced cancers in a personalized manner [16]. Indeed, the cancerous manifestations are closely related to the synthesis pathways of inflammatory proteins called cytokines.

It seems that the analysis of the synthesis rate of peptide segments by new sequencing technologies is a tool for the future in transversal medicine and to unite bioinformatics with clinical practice. To date, experimental methods, aiming to associate bioinformatics with medical research, have identified 254 protein translation abnormalities [17, 18] (resulting from an anomaly in the spatial configuration of proteins) likely to be involved in pathologic manifestations. Twenty-five protein syntheses have been identified as biomarker potential in the specific recognition of diseases.

Therefore proteomics (exhaustive analysis of the battery of proteins expressed by a given tissue or cell population and its variations according to the physiological or pathological state) is in full expansion; it has become accessible to many laboratories, in particular through the development and modernization of mass spectrometry. In association with studies in genomics and metabolomics, it presents a link of the future between bioinformatics and clinical practice: whether in the identification of biomarkers in the real-time analysis of the evolution of diseases or in the evaluation of drug treatment [19]. Today, some databases and software (Table 6.1 segment 4 and Table 6.2) serve as references by gathering large volumes of data from mass spectrometry experiments with micro-array preparation technology according to standards of forward phase and reverse phase [20].

The public reference database for proteomics is PRIDE (Proteomics IDEntifications – http://www.ebi.ac.uk/pride). PRIDE gather and standardize all proteomic data from different species, different tissues, and different cell types. Each of the 42,000 proteins, 270,000 protein sequences, and 100 million accumulated spectrometric data are standardized and then archived [21] (Table 6.2).

All of these databases and software are very widely used in the identification and profiling of a large number of proteins. But one can go further: these data allow nowadays to understand the protein environment, by the analysis of protein–protein interactions and post-translational modifications. Indeed, by

Table 6.2 Proteomics software

Software	Use	Reference
Mascot [22]	Peptide search engine by spectrometric matching	http://www.matrixscience.com
SEQUEST [23]	Identification of the protein/peptide coupling in the protein database identified by SEQUEST mass spectrometry	https://omictools.com/sequest-tool
PLGS (*ProteinLynx Global Server*)	Allows a qualitative and quantitative analysis of proteins for proteomics research. Allows statistical filters to be applied to minimize the false-positive rate	http://www.waters.com/waters/fr_FR/ProteinLynx-Global-SERVER-%28PLGS%29/nav.htm?cid=513821&locale=fr_FR
PEAKS [24]	Allows the identification of new peptides without database, thanks to the analysis of the amino acid sequence	http://www.bioinfor.com/

identifying proteins qualitatively and quantitatively, researchers are able to explore more broadly the expression profile of proteins in addition to variations in biological systems. Moreover, the interpretation of these data still requires a special precaution because of the sensitive nature of the analyzed data but also to limit statistical errors [25]. Understanding and interpreting internal and external protein interactions require the introduction of a standard or a standardization adapted to the purpose of the research. Whether in protein synthesis pathway analysis, protein interaction analysis, or post-translational modifications, a standard must be applied. For this reason, in 2009, Riffle and Eng [26] facilitated the dissemination of proteomic data by building a public database. This approach has made it possible to establish bioinformatics standards necessary for the storage, exchange, and sharing of data from mass spectrometry research and identification of the post-translational functionalities of proteins.

6.3 MicroRNA Biomarkers (Transcriptomics)

In addition to proteomic research previously described, recognition of microRNAs (miRNAs) as potential biomarkers is intensifying. The new sequencing methods (RNA sequencing or RNA-Seq) allowed a very precise exploration of the RNA sequences and to map the transcriptome complex. The latter is a reflection of cellular activity. It allows to discover many non-coding RNA sequences such as miRNA (microRNA), siRNA (short interfering RNA), piRNA (piwi-interacting RNA), and lncRNA (long non-coding RNA) involved in epigenetic mechanisms. These RNA sequences are known to be non-coding and intervene in the regulation of gene expression at transcriptional and post-transcriptional stages. Non-coding RNA sequences such as miRNAs, siRNAs, and piRNAs seem to play a specific role in the compaction state of DNA, especially in the formation of heterochromatin, which is possible, thanks to histone proteins, but also in the methylation of DNA strands and the inhibition of genetic sequence called silencing. Long RNA sequences (>200 nucleotides) can also influence gene expression at specific genome locations. The study of miRNAs has specifically led to a better understanding of the genetic mechanisms underlying diabetes or cancers [27]. In addition to the different biomarker potentials previously mentioned, databases are now dedicated to the study of miRNAs (Table 6.1 segment 5).

A good application is the diagnosis of musculoskeletal pathologies. In 2013, the discovery of a microRNA linked to the dystrophin protein made it possible to understand the evolution of Duchenne muscular dystrophy [32]. miRNAs are excellent biomarkers for the physio-pathological analysis of musculoskeletal tissue, and they allow the identification of many pathologies. In fact, dystrophy-associated miRNAs

Table 6.3 Example of miRNA biomarkers

miRNA biomarker	Associated disease	Reference
miR-494 miR-1973 miR-21	Classification of Hodgkin lymphoma	Jones et al. [33]
Keratine-18 miRNA-122	Evaluation of drug-induced liver injury	Thulin et al. [34]
miR-31 miR-206 miR-424 miR-146a	Inflammatory bowel disease	Lin et al. [35]
miR-21 miR-106b miR-17 miR-18 miR-20a	Stomach cancer	Wang et al. [36]
miRs-200 et miR-100	Ovarian epithelium cancer	Chen et al. [37]
miRNA-17 miR-18a miR-20a	Retinoblastoma	Beta et al. [38]
miR-618 miR-650	Viral hepatitis C with associated hepatocarcinoma	Abdalla and haj-Ahmad [39]
miR-1254 miR574-5p	Large cell lung cancer	Foss et al. [40]
miR-519d miR_647	Prostate cancer	Long et al. [41]

(dystromiRs) make it possible to characterize the progression of muscular pathology. Like the myogenic miR-206, which is a transcription factor of the myogenin molecule, dystromiRs have a concentration that increases when muscle tissue is degraded experimentally.

In Table 6.3, we present some relevant miRNAs as biomarkers from the Biomarkers and Systems Medicine (BSM) website [42].

Through the BSM database, we can see that the associated interpretation of miRNAs makes it possible to provide a prognosis or a diagnosis. miRNAs co-evolve in a network and some miRNAs are involved as biomarkers of different pathologies, as miR-21 present in stomach cancer or in Hodgkin's lymphoma.

6.4 Biomarkers of Circadian Rhythms

Other biomarkers are particularly useful in physiopathology, in particular to identify their evolution over time. These include molecules whose quantity changes with the rhythm of biological cycles (Table 6.4: Biomarkers of the circadian rhythm). The biological constants can evolve through the rhythms to which the organism will be subjected: annual, biannual, seasonal, monthly, or even circadian and ultradian rhythm. Most of the endogenous biological variations are due to the circadian rhythm which makes it possible to ensure the control, the synthesis, the renewal, and the degradation of the basic molecules necessary for cellular mechanisms.

The identification of biomarkers reflecting the evolution of tissues in a day is an issue of biological research to classify natural tendencies of the body during normal biological aging or during pathophysiological mechanisms.

Molecules evolving at the rate of the biological clock can identify upstream pathophysiological phenomena, such as breast cancer or psychiatric disorders. These molecules are intimately related to hormonal mechanisms [49].

Table 6.4 Circadian rhythm biomarkers

Circadian biomarker	Variation	Disease	Reference
Melatonin	Decrease	Bipolar disease	Milhiet et al. [43]
Neuronal PAS domain protein-2 (NPAS2)	Expression increase	Breast cancer	Yi et al. [44]
Period circadian protein homolog 3 protein (PER3)	Decrease	Breast cancer	Zhu et al. [45]
Surface body temperature variation	Dysregulation	Cancer	Scully et al. [46]
Salivary melatonin, daily cortisol levels	Decrease	Metabolic syndrome	Corbalan-Tutau et al. [47]
Cytokines, cortisol, and inflammatory markers (TNF, VCAM)	Increase	Multiple sclerosis	Wipfler et al. [48]

6.5 Biomarkers of Inflammation

In recent years, a strong link between the microbiota, the digestive tract, and the brain has been highlighted. This link is undoubtedly a major avenue of exploration in biomedical research in the years to come [50].

This underlies endocrine, autonomic, immunological, and metabolic mechanisms. This combination of physiologies makes it possible to protect the nourishing pathways of the body by chemical feedback to ensure the body's internal homeostasis in the external environment [51]. New studies have identified biomarkers of low back pain. Low back pain is the leading cause of disability, caused by a variety of spinal disorders, including intervertebral disc degeneration, disc herniation, spinal stenosis, and facet arthritis. The discovery of these new biomarkers is crucial in understanding everyday pain.

References

1. Markman BT. Biomarkers for go/no go decisions. In: Lenz H-J, editor. Biomarkers in oncology – prediction and prognosis. New York: Springer; 2013.
2. Johnson DS, et al. Genome-wide mapping of in vivo protein-DNA interactions. Science. 2007;316(5830):1497–502.
3. Online Mendelian Inheritance in Man (OMIM). 2018. Copyright® 1966–2018 Johns Hopkins University.
4. Guigo R, et al. Prediction of gene structure. J Mol Biol. 1992;226(1):141–57.
5. Davies KE, et al. Linkage analysis of two cloned DNA sequences flanking the Duchenne muscular dystrophy locus on the short arm of the human X chromosome. Nucleic Acids Res. 1983;11(8):2303–12.
6. Donis-Keller H, et al. A genetic linkage map of the human genome. Cell. 1987;51(2):319–37.
7. Catalog of published genome-wide association studies. 2018. Available from: https://www.genome.gov/26525384/catalog-of-published-genomewide-association-studies/.
8. Hebbring SJ. The challenges, advantages and future of phenome-wide association studies. Immunology. 2014;141(2):157–65.
9. SNOMED CT. http://www.snomed.com. 2018.
10. Ontobee. 2018. Available from: http://www.ontobee.org/.

11. Human phenotype ontology. 2018. Available from: http://human-phenotype-ontology.github.io.
12. Diseases database ver 2.0; Medical lists and links diseases database. 2018. Available from: http://www.diseasesdatabase.com/.
13. HIPAA administrative simplification. American Medical Association. 2018.
14. Pitteri S, Hanash S. A systems approach to the proteomic identification of novel cancer biomarkers – IOS Press. Dis Mark. 2010;28(4).
15. Hathout Y, et al. Clinical utility of serum biomarkers in Duchenne muscular dystrophy. Clin Proteomics. 2016;13:9.
16. Pellegrini P, et al. Gender-specific cytokine pathways, targets, and biomarkers for the switch from health to adenoma and colorectal cancer. Clin Dev Immunol. 2011;2011:819724.
17. Huang Y, Zhu H. Protein array-based approaches for biomarker discovery in cancer. Genom Proteom Bioinform. 2017;15(2):73–81.
18. Tu S, et al. Protein microarrays for studies of drug mechanisms and biomarker discovery in the era of systems biology. Curr Pharm Des. 2014;20(1):49–55.
19. Yan Q. Translational bioinformatics and systems biology methods for personalized medicine. 1st Ed. Published: 20th April 2017 Academic Press eBook. ISBN:9780128043882.
20. Chandra H, Reddy PJ, Srivastava S. Protein microarrays and novel detection platforms. Expert Rev Proteomics. 2011;8(1):61–79.
21. Perez-Riverol Y, et al. Making proteomics data accessible and reusable: current state of proteomics databases and repositories. Proteomics. 2015;15(5–6):930–49.
22. Koenig T, et al. Robust prediction of the MASCOT score for an improved quality assessment in mass spectrometric proteomics. J Proteome Res. 2008;7(9):3708–17.
23. Diament BJ, Noble WS. Faster SEQUEST searching for peptide identification from tandem mass spectra. 2011.
24. Ma B, et al. PEAKS: powerful software for peptide de novo sequencing by tandem mass spectrometry. Rapid Commun Mass Spectrom. 2003;17(20):2337–42.
25. Listgarten J, Emili A. Statistical and computational methods for comparative proteomic profiling using liquid chromatography-tandem mass spectrometry. Mol Cell Proteomics. 2005;4(4):419–34.
26. Riffle M, Eng JK. Proteomics data repositories. Proteomics. 2009;9(20):4653–63.
27. Pescador N, et al. Serum circulating microRNA profiling for identification of potential type 2 diabetes and obesity biomarkers. PLoS One. 2013;8(10):e77251.
28. Kozomara A, Griffiths-Jones S. miRBase: annotating high confidence microRNAs using deep sequencing data. Nucleic Acids Res. 2014;42:D68–73.
29. Andres-Leon E, et al. miRGate: a curated database of human, mouse and rat miRNA-mRNA targets. Database (Oxford). 2015;2015:bav035.
30. Chou CH, et al. miRTarBase 2016: updates to the experimentally validated miRNA-target interactions database. Nucleic Acids Res. 2016;44(D1):D239–47.
31. Wong N, Wang X. miRDB: an online resource for microRNA target prediction and functional annotations. Nucleic Acids Res. 2015;43(Database issue):D146–52.
32. Roberts TC, et al. Extracellular microRNAs are dynamic non-vesicular biomarkers of muscle turnover. Nucleic Acids Res. 2013;41:9500–13.
33. Jones K, et al. Plasma microRNA are disease response biomarkers in classical Hodgkin lymphoma. Clin Cancer Res. 2014;20(1):253–64.
34. Thulin P, et al. Keratin-18 and microRNA-122 complement alanine aminotransferase as novel safety biomarkers for drug-induced liver injury in two human cohorts. Liver Int. 2014;34(3):367–78.
35. Lin J, et al. Novel specific microRNA biomarkers in idiopathic inflammatory bowel disease unrelated to disease activity. Mod Pathol. 2014;27(4):602–8.
36. Wang JL, et al. Candidate microRNA biomarkers in human gastric cancer: a systematic review and validation study. PLoS One. 2013;8(9):e73683.

37. Chen Y, Zhang L, Hao Q. Candidate microRNA biomarkers in human epithelial ovarian cancer: systematic review profiling studies and experimental validation. Cancer Cell Int. 2013;13:86.
38. Beta M, et al. Identification and insilico analysis of retinoblastoma serum microRNA profile and gene targets towards prediction of novel serum biomarkers. Bioinform Biol Insights. 2013;7:21–34.
39. Abdalla MA, Haj-Ahmad Y. Promising candidate urinary microRNA biomarkers for the early detection of hepatocellular carcinoma among high-risk hepatitis C virus Egyptian patients. J Cancer. 2012;3:19–31.
40. Foss KM, et al. miR-1254 and miR-574-5p: serum-based microRNA biomarkers for early-stage non-small cell lung cancer. J Thorac Oncol. 2011;6(3):482–8.
41. Long Q, et al. Protein-coding and microRNA biomarkers of recurrence of prostate cancer following radical prostatectomy. Am J Pathol. 2011;179(1):46–54.
42. BSM. Biomarkers and systems medicine. 2017. Available from: http://pharmtao.com/health/category/systems-medicine/biomarkers-systems-medicine.
43. Milhiet V, et al. Circadian biomarkers, circadian genes and bipolar disorders. J Physiol Paris. 2011;105(4–6):183–9.
44. Yi C, et al. The circadian gene NPAS2 is a novel prognostic biomarker for breast cancer. Breast Cancer Res Treat. 2010;120(3):663–9.
45. Zhu Y, et al. Period3 structural variation: a circadian biomarker associated with breast cancer in young women. Cancer Epidemiol Biomark Prev. 2005;14(1):268–70.
46. Scully CG, et al. Skin surface temperature rhythms as potential circadian biomarkers for personalized chronotherapeutics in cancer patients. Interface Focus. 2011;1:48–60.
47. Corbalan-Tutau D, et al. Daily profile in two circadian markers "melatonin and cortisol" and associations with metabolic syndrome components. Physiol Behav. 2014;123:231–5.
48. Wipfler P, et al. Circadian rhythmicity of inflammatory serum parameters: a neglected issue in the search of biomarkers in multiple sclerosis. J Neurol. 2013;260(1):221–7.
49. Rattan S. Hormones in ageing and longevity. Springer; 2018.
50. Sherwin E, et al. May the force be with you: the light and dark sides of the microbiota-gut-brain axis in neuropsychiatry. CNS Drugs. 2016;30(11):1019–41.
51. Sandhu KV, et al. Feeding the microbiota-gut-brain axis: diet, microbiome, and neuropsychiatry. Transl Res. 2017;179:223–44.

Chapter 7
Regenerative Medicine in the Digital Age

Arthur André

Regenerative medicine deals with the "process of replacing, engineering or regenerating human cells, tissues or organs to restore or establish normal function" [1]. The development of information technology (IT) has allowed great advancements in this field of research with gene editing, signal transformation algorithms, 3D bioprinting, and anti-aging medicine. Therefore, stem cell therapy, gene therapy, human–machine interface, and 3D organ printing—all different aspects of regenerative medicine—are growing rapidly, thanks to emerging technologies.

7.1 Stem-Cell Therapy

Most tissues in the body have some ability to recover from injury and progression of disease. However this ability of self-repair can be exceeded and tissue function compromised. Stem cells are endorsed with indefinite cell division potential, can transdifferentiate into other types of cells, and have emerged as frontline regenerative medicine source in recent times. Stem-Cell therapy principles are already used in daily practice: bone marrow transplant has been used for decades to treat some kind of leukemia and lymphoma. Nowadays, advancements in gene editing and tissue engineering technology have endorsed the ex vivo remodeling of stem cells grown into 3D organoids and tissue structures for personalized applications [2].

This has generated great hope in the use of these cells to increase this repair capacity and has led to successes in regenerative medicine, particularly concerning muscle and cardiac tissues [3, 4]. Promising results are also presented for diabetes [5], retinopathy [6], liver failure [7], spinal cord injury [8], and lupus [9]. The idea

A. André (✉)
Pitié-Salpêtrière Hospital, Assistance Publique Hôpitaux de Paris, Paris, France

Sorbonne Université, Paris, France
e-mail: arthur.andre@neurochirurgie.fr

© Springer Nature Switzerland AG 2019

A. André (ed.), *Digital Medicine*, Health Informatics,
https://doi.org/10.1007/978-3-319-98216-8_7

is then to partially or completely repair the cellular elements and their circuits. The other option of cell therapy is to mobilize stem cell niches in the body to recruit endogenous stem/progenitor cells, relying on "remaining" circuits.

Embryonic stem cells (ESCs) can be transdifferentiated into any kinds of cells representing three germ layers of the body, representing one of the most promising source of regenerative medicine [2]. ESCs are pluripotent in their nature and can potentially give any kind of human tissue. Ectopic expression or functional restoration of endogenous pluripotency factors epigenetically transforms terminally differentiated cells into ESCs-like cells known as induced pluripotent stem cells (iPSCs) [10]. In 2017, a Japanese man was the first to be treated by reprogrammed cells for macular degeneration [6, 11].

Stem cells can be categorized as embryonic stem cells (ESCs), induced pluripotent stem cells (iPSCs), tissue-specific progenitor stem cells (TSPSCs), mesenchymal stem cells (MSCs), umbilical cord stem cells (UCSCs), and bone marrow stem cells (BMSCs). Ethical concerns limit the applications of ESCs, where set guidelines need to be followed; in that case TSPSCs, MSCs, UCSCs, BMSCs, and iPSCs are often explored as alternatives [2].

7.2 Gene Therapy

Gene therapy is the delivery of a gene for therapeutic purposes in the cells of an organism, thanks to a vector, most often of viral origin. In the future, this technique may allow doctors to treat a disorder by inserting a gene into a patient's cells instead of using drugs or surgery.

Gene therapy may be classified into two types:

- Germline gene therapy (GGT): Introduction by transfer of a therapeutic gene into a very early embryo, even unicellular. The gene is transmitted to all cells, including gametes (thus to the offspring). The only present applications are in animals (transgenesis): animal models of human diseases or production of animal substances of biological interest. Australia, Canada, France, Germany, Israel, Switzerland, and the Netherlands prohibit the application of GGT to human beings, for technical and ethical reasons, including embryo sorting, eugenics, and insufficient knowledge about possible risks to future generations.
- Somatic gene therapy: Introduction by transfer of a therapeutic gene into the somatic cells of a constituted organism, without modifying its heredity.

Several approaches to somatic gene therapy exist, including:

- Replacing a mutated gene that causes disease with a healthy copy of the gene. Validated examples of this approach include the treatment of severe combined X-linked immunodeficiency [12] and adrenoleukodystrophy [13]. These diseases lend themselves particularly well to gene therapy, as they can be treated by easily accessible (hematopoietic) stem cell populations that can be transduced

ex vivo (allowing greater transduction efficiency) with a single gene and expanded in vivo.
- Inactivating, or "knocking out," a mutated gene that is functioning improperly.
- Introducing a new gene into the body to help fight a disease. This last option is a promising treatment for multigenic diseases, neurodegenerative diseases, some types of cancer, and even certain viral infections.

Over 700 clinical trials utilizing somatic gene therapy are underway in the world. The complete correction of a genetic disorder or the replacement of multiple genes is not yet possible. Only a few of the trials are in the advanced stages [14]. After a lot of proof-of-concept works in animal models, the last decade has been marked by the success of phase I/II gene therapy trials in different areas: the blood system (immune deficiency, hemoglobinopathies), muscle (Duchenne muscular dystrophy), liver (hemophilia), and brain (lysosomal diseases, adrenoleukodystrophy) [12, 13, 15].

7.2.1 Gene Editing and CRISPR /Cas9 Technology

Genome correction techniques include genetic engineering techniques in which one or more pieces of DNA are inserted, replaced, or removed from a genome using artificially modified nucleases (so-called molecular scissors).

Four families of modified nucleases have been developed in genetic engineering:

- Zinc finger nucleases (ZFNs)
- Transcription activator-like effector nucleases (TALENs)
- Meganuclease (engineered meganuclease, re-engineered homing endonucleases) engineered by genetic engineering to recognize and target specific genome sequences
- Clustered regularly interspaced short palindromic repeats or CRISPR/Cas system, probably the most promising technique to date.

CRISPR are a series of short direct repeats (21 to 37 bp) spaced by unique sequences of 20 to 40 bp, present in many bacteria and almost all archaea, and useful in adaptive immune system recognition and defense of viral damage. *Cas9* (CRISPR-associated protein 9) is an RNA-guided DNA nuclease enzyme. The location at which Cas9 cuts DNA is specified by a guide RNA comprised of a crRNA component and a tracrRNA component, either individually or combined together as a single guide RNA (gRNA). Once the molecular scissors make a cut in the DNA, the cell's own robust natural repair machinery repairs the cut, a process that can disrupt or delete a disease-causing gene, or correct that gene if the desired DNA is added as a template.

The CRISPR-Cas9 system, developed in particular by researchers Emmanuelle Charpentier and Jennifer Doudna, has since 2010 become a revolutionary genetic engineering tool that makes it possible to modify DNA sequences more easily and more precisely. It may ultimately help to eliminate some diseases but raises questions on medical and environmental ethics with respect to eugenics and the environmental consequences of genome manipulation, when applied to hereditary cells.

7.3 The Example of Neurodegenerative Diseases

Although the existence of neurogenesis in the adult brain is now universally accepted by the scientific community, it has long been difficult to refute the longstanding dogma that new neurons cannot be generated in the adult brain. In addition, the elucidation of neurogenesis steps and their functional significance, as well as the molecular processes underlying these events, have only recently been enabled by laboratory-based technological advances [16].

Indeed, stem/progenitor cells can be isolated from neurogenic regions and can then be grafted into the neurogenic environment where they have the ability to differentiate into neurons in the adult rat [17, 18]. Although there has been very promising evidence of functional integration of transplanted glial progenitor cells in a mouse model of demyelinating disease [19], transplantation of adult neural progenitor cells derived in a neurogenic region of brain outcome after birth shows the lack of commitment to a specific neuronal lineage of these cells [20, 21]. As a result, protocols that have transplanted stem cells into the brain to evaluate therapeutic potential have either investigated the incorporation of exogenous stem cells into neurogenic regions to restore normal function, as in temporal epilepsy [22], or used stem cells as cellular vectors to support dysfunctional circuits, for example, in the treatment of Parkinson's disease (PD), by attempting to restore dopaminergic production [23, 24]. Thus, several fetal neural transplantation trials have shown promising results by improving the motor score of Parkinsonian patients [25–27], although dyskinesias (a type of movement disorder) persisted in majority of PD patients in these studies. That is why the European multicenter trial TransEURO, currently underway, is trying to prove the effectiveness of a fetal cell transplant with a stricter selection of patients.

Neural grafts have also been attempted in humans with Huntington's disease in the striatum. A pilot French trial in five patients showed a benefit at 2 years in three of them, although the motor and cognitive deterioration according to the natural history of the disease was not stopped at 6 years of follow-up [28].

Progress in gene therapy has allowed for novel treatments of both genetic and acquired neurodegenerative diseases such as lysosomal storage disorders, Alzheimer's disease (AD), Parkinson's disease (PD), Huntington's disease (HD), amyotrophic lateral sclerosis (ALS), and spinal muscular atrophy (SMA) [29].

One of the major signal classes studied for both repairing diseased neurons and supporting stem cell proliferation are trophic factors. Neurotrophins are a class of agents that include Nerve Growth Factor (NGF), Brain-Derived Neurotrophic Factor (BDNF), and neurotrophin-3 (NT3). Basic Fibroblast Growth Factor (bFGF) and Vascular Endothelial Growth Factor (VEGF) are also important pro-neurogenic growth factors [30, 31]. For example, adeno-associated, virus-mediated BDNF delivery alone enhances the recruitment of rat urinary progenitor (neuroblastic) cells and promotes neuronal differentiation [32, 33].

Thus the genetic manipulation of stem cells is being developed for major neurodegenerative diseases of the central nervous system. The use of NGF in the treatment of AD is an excellent example of neurotrophic factor gene therapy. A Phase I clinical trial was conducted using autologous fibroblasts with the transduced NGF gene delivered to the cholinergic structures of the forebrain. This resulted in a reduction in the rate of cognitive decline without any adverse effects. Autopsy showed robust dendritic growth in a treated individual [34]. In another study, bFGF/BDNF viral vector-mediated expression in the hippocampus of an epileptic lesion model increased neurogenesis and reduced neuronal lesions [35]. In a stroke animal model, the VEGF gene transfer has been found to increase neurogenesis, recruit stem cells for infarction, and reduce infarct size [36]. In an example of ex vivo genetic manipulation of the central nervous system, Gu et al. generated stem cells expressing NT3 and were able to transplant them to parkinsonian rats with positive results on motor symptoms.

Concerning PD, several Phase I/II gene therapy trials have been published to date with positive results: the GAD study which aims to transfer a precursor of glutamic acid decarboxylase (which is involved in GABA metabolism) in the subthalamic nucleus [37]; the CERE-120 study which aims to administer a neurotrophic factor, the Glial cell line-Derived Neurotrophic Factor (GDNF), in the striatum [38]; or the ProSavin study [39].

7.4 Brain-Computer Interface

Brain–computer interfaces (BCI) are born to control "action by thought" to restore the capabilities of patients whose functional connectivity has been altered between the central and peripheral nervous systems [40].

Over the last 15 years, the BCI and, among them, neuroprostheses have largely been developed with the aim of improving the quality of life of patients affected by motor and cognitive neurological deficits [41].

However, to date, these advances are insufficient to hope to recreate a specific function without a thorough knowledge of the underlying neural network [42]. Indeed the motor and cognitive function can be subdivided into spatially distinct elementary subtasks in the white substance fiber networks [43]. In this view, direct electrical stimulation, applied during low-grade glioma awake surgery, provides an unexpected new description of this functional connectivity [42]. Since the initial work of Penfield, which identified the homunculus corresponding to the primary motor area (M1) [44], the cortical electrical stimulation allowed to identify with precision the circuits of the motricity and the language in particular with the prospect to build BCI.

Thus Hochberg et al. have proposed a cortical array of microelectrodes composed of 96 channels implanted on M1 allowing a quadriplegic patient to control a robotic arm [45], although the precise control of the arm by the patient was not optimal. This last observation shows that the execution of the voluntary movement also depends on higher level networks [41].

For example, we know the activation of M1 without movement in response to a movement performed by someone else: this is the phenomenon of mirror neurons, identified in non-human primates and then in humans [46]. It has been proposed that this ability of the motor cortex to be dissociated from voluntary movement can be used in the control of BCI [47].

In addition, the identification of "negative motor areas," i.e., blocking motion during stimulation, located in the prefrontal cortex (anterior to supplementary motor area, beyond M1), connected with the parietal cortex [48], helps to understand how the brain estimates the state of motion and corrects it in real time. These areas appear as other potential targets for BCI recording [42].

It is also about restoring sensory abilities. Recent improvements in cochlear or retinal implants also represent substantial advances in BCI.

Finally, recent research suggests that there is a possibility to restore cognitive functions. Thus, a hippocampal implant has allowed a rat to improve his memory [49], or and registrations in the prefrontal cortex of the primate could held to basic behavioral control [50].

7.4.1 In Vitro Models

The realization of such devices involves knowing and decrypting the intrinsic neuronal signal and then reinjecting it to direct it toward a specific goal. In basic research, many advances have been made in vitro that make it possible to imagine what will be the BCI of the future. Bonifazi et al. have proposed through the Brain Bow project a "bottom-up" approach to neuroprostheses [40]: cell cultures of neurons make it possible to define a circuit of finite and identifiable connections, transferred to an in silico neural network, which implements a chip "Neuromorphic" according to the name of authors. Thus a dozen of GABAergic neurons have been identified as having a spontaneous, synchronous, and collective burst activity (oscillatory activity in the gamma-theta range, with the frequency of burst increasing with the number of neurons). The decoding of this activity, then its re-encoding according to computational models [51], thus makes it possible to create a true neural prosthesis capable of replicating the function of an injured neural circuit without replacing all its architecture.

Tessadori et al. presented their project Hybrain 2, which consists of coupling a culture of hippocampal cells with a robot equipped with sensory and wheel sensors. The hippocampal cells were cultured on a multi-electrode array, and the team managed to achieve bidirectional control of the neuro-robotics system, with a neuron-controlled robot on an obstacle course, and reinforcement by learning the tasks of shifting [52, 53]. Similarly, Brewer et al. reconstructed a hippocampal trisynaptic loop in vitro on an MEA with small tunnels for neural growth [53].

7.4.2 Computational Models

In the field of control of the neuronal message, several theoretical models of simulation have been proposed. Indeed, the increasing computing power of computers makes it possible to model. The classical model considered the optimal stimulus as the most effective stimulus to excite a neuronal population, as if there was a linear relationship between the sensory stimulus and the neuronal response. This becomes more complex when the population in question receives several signals, even more if one considers several neuronal layers. If the neural architecture is not precisely known a priori, as is often the case, closed-loop computational models have been proposed: thus, the optimal stimulus is said to be "adaptive" in real time. Depending on the response recorded continuously to a stimulus presented, the computer can optimize the stimulus in real time in order to have a stronger response. Thus the computer can learn to adapt to a specific neuronal population; for example, Hanuschkin et al. modelized the sensorimotor loop by which birds produce stereotyped songs [54]. Skocik and Kozhevnikov [55] have similarly presented a real-time audio feedback system to study song learning by birds. Molkov et al. modeled the roles of local and (brain stem) and distal (lung) feedback in mammalian respiratory control [56].

7.4.3 Animal Models and Applications in Humans

BCI has also been developed in applied research in animals. Thus, in monkeys, a connected prosthesis could be controlled, thanks to the unitary registration of the neurons of motor cortex in rats [57] and in primates [58, 59]. Moreover, the simultaneous recording of several cortical neuron populations (primary motor cortex, premotor, and posterior parietal), still in the monkey, allowed the real-time prediction of the movement, thanks to the transformation of the neural signal by a linear algorithm, the adapted control of a robotic arm [60].

In humans, there are already reported cases of prostheses controlled by a brain–machine interface [45, 61]. Moreover, electrocorticogram recordings in epileptic patients have made it possible to identify specific patterns related to a specific action [62]. Fernandez-Vargas et al. [63] studied closed-loop optimization of a flickering light screen as part of a visual evoked potential, which could be used by locked-in patients to communicate. Communication has remained impossible for persons suffering from complete motor paralysis but intact cognitive and emotional processing, such as post-stroke or amyotrophic lateral sclerosis patients, and BCI could be a very promising regenerative tool [64, 65]. Concerning motor recovery and potential link to exoskeleton, BCIs also have great present and future potential [66]. Finally, some research teams have made this last decade tremendous advances concerning hippocampal or enthorinal prothesis and enhanced memory, which is no longer science-fiction [67–69].

7.5 3D Printing

Three-dimensional (3D) bioprinting technology promises to bridge the divergence between artificially engineered tissue constructs and native tissues [70, 71].

We recently moved from using synthetic implants and tissue grafts to a tissue engineering approach that uses degradable porous material scaffolds integrated with biological cells or molecules to regenerate tissues [70, 72].

Current 3D printing systems build objects through several techniques: photopolymerized liquid monomer, sintered powdered materials, processed material, either thermally or chemically, as it passes through a nozzle, or print materials, such as chemical binder onto powder. In printing 3D live tissues, recent research focuses on developing technology to simultaneously deposit hydrogels with live cells to form 3D tissue structures [71].

Many trials have been made concerning different tissues [73]:

- Blood vessels, developed by researchers at Harvard University [74]. The vasculature network they created enables fluids, nutrients, and cell growth factors to navigate uniformly throughout the artificial vessel tissue.
- Bones: Professor Susmita Bose of Washington State University used a 3D printer to bind chemicals to a ceramic powder, creating intricate ceramic scaffolds that promote the growth of the bone in any shape [75, 76]. It could help hip and knee replacements.
- Heart valve: Researchers from Cornell University have 3D-printed a heart valve possessing the same anatomical architecture as the original valve. It is presently being tested in big animals. He used a combination of cells and biomaterials to control the valve's stiffness [77].
- Replicating human ears: L. Bonassar from Cornell University used 3D photos of human ears to create ear molds [78]. The molds were then filled with a gel containing bovine cartilage cells suspended in collagen, which held the shape of the ear while cells grew their extracellular matrix.
- Synthetic skin is being developed at the Wake Forest Institute for Regenerative Medicine [79] or at the Burns Institute at the Southwest Hospital in Chongqing [80].

Synthetic organs: The company Organovo has already bioprinted liver and kidney tissues since 2014. Their next step is to print liver parts for transplantation. Bioprinted livers could also be used in the pharmaceutical industry to replace animal models for analyzing the toxicity of new drugs.

7.6 Anti-aging Medicine

The phenotype of aging is described by a process of physical and cognitive decline, which ultimately results in "fragility." Fragility is a clinical concept defined as a biological syndrome in which a cumulative decline in the spare capacity of multiple biological systems results in an abnormal vulnerability to common stress factors [81]. Fragility accompanies the rise of global inflammation associated with aging, or "inflamm-ageing" [82].

Aging is a real challenge for bioinformatics, which aims to conceptualize the evolution of health during normal aging. The latter is present at all scales: genetic, molecular, cellular, and phenotypic.

Current technological advances make it possible, by combining genetic, proteomic, and clinical data, to analyze individually certain diseases of aging. Aging is systematically linked to cellular and molecular ubiquitous alterations, which are biomarkers of aging, and which a largely consensual review, published in Cell, classified into nine 314 categories [83]:

- Cellular senescence
- Telomere attrition
- Mitochondrial dysfunction
- Genomic instability
- Epigenetic alterations
- Stem cell exhaustion
- Altered intercellular communications
- Deregulated nutrient sensing
- Loss of proteostasis

Aging is a very important field of biomedical research. To date there is a large amount of data from new data storage technologies as described above in genomics, proteomics, microRNA (miRNA), etc.

However, none of this data is currently used for diagnostic purposes. It is still difficult to extract potential biomarkers in correlation with the entire pathophysiology of aging. Several databases are specifically dedicated to the census of aging biomarkers:

- GenAge: Gene of longevity and aging [84]
- Gene Aging Nexus: Database of the research community in biology and geriatrics [85]
- AgeID: Genes of aging [86]
- MARK-Age database: Genomics, proteomics, and miRNA of aging with diseases associated with the chronology of aging [87].

References

1. Mason C, Dunnill P. A brief definition of regenerative medicine. Regen Med. 2008;3(1):1–5.
2. Mahla RS. Stem cells applications in regenerative medicine and disease therapeutics. Int J Cell Biol. 2016;2016:1.
3. Cerletti M, et al. Highly efficient, functional engraftment of skeletal muscle stem cells in dystrophic muscles. Cell. 2008;134(1):37–47.
4. Ghodsizad A, et al. Transplanted human cord blood-derived unrestricted somatic stem cells improve left-ventricular function and prevent left-ventricular dilation and scar formation after acute myocardial infarction. Heart. 2009;95(1):27–35.

5. Bruin J, et al. Treating diet-induced diabetes and obesity with human embryonic stem cell-derived pancreatic progenitor cells and antidiabetic drugs. Stem Cell Rep. 2015;4(4):605–20.
6. Mandai M, et al. Autologous induced stem-cell–derived retinal cells for macular degeneration. N Engl J Med. 2017;376(11):1038–46. https://doi.org/10.1056/NEJMoa1608368.
7. Cameron K, et al. Serum-free directed differentiation of human embryonic stem cells to hepatocytes. Methods Mol Biol. 2015;1250:105–11.
8. Shroff G, Gupta R. Human embryonic stem cells in the treatment of patients with spinal cord injury. Ann Neurosci. 2015;22(4):208–16.
9. Sun L, et al. Umbilical cord mesenchymal stem cell transplantation in severe and refractory systemic lupus erythematosus. Arthritis Rheum. 2010;62(8):2467–75.
10. Takahashi K, Yamanaka S. Induction of pluripotent stem cells from mouse embryonic and adult fibroblast cultures by defined factors. Cell. 2006;126(4):663–76.
11. Cyranoski D. Japanese man is first to receive 'reprogrammed' stem cells from another person. Nature News. 2018.
12. Hacein-Bey-Abina S, et al. Efficacy of gene therapy for X-linked severe combined immunodeficiency. N Engl J Med. 2010;363(4):355–64.
13. Cartier N, et al. Hematopoietic stem cell gene therapy with a lentiviral vector in X-linked adrenoleukodystrophy. Science. 2009;326(5954):818–23.
14. Mavilio F, Ferrari G. Genetic modification of somatic stem cells: the progress, problems and prospects of a new therapeutic technology. EMBO Rep. 2008;9:S64–9.
15. Naldini L. Medicine. A comeback for gene therapy. Science. 2009;326(5954):805–6.
16. Kempermann G, et al. Milestones of neuronal development in the adult hippocampus. Trends Neurosci. 2004;27(8):447–52.
17. Gage FH, et al. Survival and differentiation of adult neuronal progenitor cells transplanted to the adult brain. Proc Natl Acad Sci U S A. 1995;92(25):11879–83.
18. Suhonen JO, et al. Differentiation of adult hippocampus-derived progenitors into olfactory neurons in vivo. Nature. 1996;383(6601):624–7.
19. Windrem MS, et al. Neonatal chimerization with human glial progenitor cells can both remyelinate and rescue the otherwise lethally hypomyelinated shiverer mouse. Cell Stem Cell. 2008;2(6):553–65.
20. Sheen VL, et al. Neural precursor differentiation following transplantation into neocortex is dependent on intrinsic developmental state and receptor competence. Exp Neurol. 1999;158(1):47–62.
21. Shihabuddin LS, et al. Adult spinal cord stem cells generate neurons after transplantation in the adult dentate gyrus. J Neurosci. 2000;20(23):8727–35.
22. Shetty AK, Hattiangady B. Concise review: prospects of stem cell therapy for temporal lobe epilepsy. Stem Cells. 2007;25(10):2396–407.
23. Freed CR, et al. Transplantation of embryonic dopamine neurons for severe Parkinson's disease. N Engl J Med. 2001;344(10):710–9.
24. Lindvall O, Kokaia Z. Prospects of stem cell therapy for replacing dopamine neurons in Parkinson's disease. Trends Pharmacol Sci. 2009;30(5):260–7.
25. Olanow CW, Freeman T, Kordower J. Transplantation of embryonic dopamine neurons for severe Parkinson's disease. N Engl J Med. 2001;345(2):146; author reply 147.
26. Dyson SC, Barker RA. Cell-based therapies for Parkinson's disease. Expert Rev Neurother. 2011;11(6):831–44.
27. Barker RA, et al. Fetal dopaminergic transplantation trials and the future of neural grafting in Parkinson's disease. Lancet Neurol. 2013;12(1):84–91.
28. Bachoud-Lévi AC, et al. Effect of fetal neural transplants in patients with Huntington's disease 6 years after surgery: a long-term follow-up study. Lancet Neurol. 2006;5(4):303–9.
29. Hocquemiller M, et al. Adeno-associated virus-based gene therapy for CNS diseases. Hum Gene Ther. 2016;27(7):478–96.
30. Temple S, Qian X. bFGF, neurotrophins, and the control or cortical neurogenesis. Neuron. 1995;15(2):249–52.

31. Jin K, et al. Vascular endothelial growth factor (VEGF) stimulates neurogenesis in vitro and in vivo. Proc Natl Acad Sci U S A. 2002;99(18):11946–50.
32. Chmielnicki E, et al. Adenovirally expressed noggin and brain-derived neurotrophic factor cooperate to induce new medium spiny neurons from resident progenitor cells in the adult striatal ventricular zone. J Neurosci. 2004;24(9):2133–42.
33. Henry RA, Hughes SM, Connor B. AAV-mediated delivery of BDNF augments neurogenesis in the normal and quinolinic acid-lesioned adult rat brain. Eur J Neurosci. 2007;25(12):3513–25.
34. Tuszynski MH, et al. A phase 1 clinical trial of nerve growth factor gene therapy for Alzheimer disease. Nat Med. 2005;11(5):551–5.
35. Paradiso B, et al. Localized delivery of fibroblast growth factor-2 and brain-derived neurotrophic factor reduces spontaneous seizures in an epilepsy model. Proc Natl Acad Sci U S A. 2009;106(17):7191–6.
36. Li SF, et al. Recombinant adeno-associated virus serotype 1-vascular endothelial growth factor promotes neurogenesis and neuromigration in the subventricular zone and rescues neuronal function in ischemic rats. Neurosurgery. 2009;65(4):771–9; discussion 779.
37. LeWitt PA, et al. AAV2-GAD gene therapy for advanced Parkinson's disease: a double-blind, sham-surgery controlled, randomised trial. Lancet Neurol. 2011;10(4):309–19.
38. Herzog CD, et al. Enhanced neurotrophic distribution, cell signaling and neuroprotection following substantia nigral versus striatal delivery of AAV2-NRTN (CERE-120). Neurobiol Dis. 2013;58:38–48.
39. Palfi S, et al. Long-term safety and tolerability of ProSavin, a lentiviral vector-based gene therapy for Parkinson's disease: a dose escalation, open-label, phase 1/2 trial. Lancet. 2014;383(9923):1138–46.
40. Bonifazi P, et al. In vitro large-scale experimental and theoretical studies for the realization of bi-directional brain-prostheses. Front Neural Circuits. 2013;7:40.
41. Lebedev MA, Nicolelis MA. Brain-machine interfaces: past, present and future. Trends Neurosci. 2006;29(9):536–46.
42. Mandonnet E, Duffau H. Understanding entangled cerebral networks: a prerequisite for restoring brain function with brain-computer interfaces. Front Syst Neurosci. 2014;8:82.
43. Shadmehr R, Krakauer JW. A computational neuroanatomy for motor control. Exp Brain Res. 2008;185(3):359–81.
44. Penfield W. Mechanisms of voluntary movement. Brain. 1954;77(1):1–17.
45. Hochberg LR, et al. Reach and grasp by people with tetraplegia using a neurally controlled robotic arm. Nature. 2012;485(7398):372–5.
46. Iacoboni M. Neural mechanisms of imitation. Curr Opin Neurobiol. 2005;15(6):632–7.
47. Schieber MH. Dissociating motor cortex from the motor. J Physiol. 2011;589(Pt 23):5613–24.
48. Mikuni N, et al. Evidence for a wide distribution of negative motor areas in the perirolandic cortex. Clin Neurophysiol. 2006;117(1):33–40.
49. Berger TW, Song D, Chan RHM, Marmarelis VZ, LaCoss J, Wills J, Hampson RE, Deadwyler SA, Granacki JJ. A hippocampal cognitive prosthesis: multi-input, multi-output nonlinear modeling and VLSI implementation. IEEE Trans Neural Syst Rehabil Eng. 20(2):198–211.
50. Opris I, et al. Closing the loop in primate prefrontal cortex: inter-laminar processing. Front Neural Circuits. 2012;6:88.
51. Izhikevich EM. Hybrid spiking models. Philos Trans A Math Phys Eng Sci. 2010;368(1930):5061–70.
52. Tessadori J, et al. Modular neuronal assemblies embodied in a closed-loop environment: toward future integration of brains and machines. Front Neural Circuits. 2012;6:99.
53. Brewer GJ, et al. Toward a self-wired active reconstruction of the hippocampal trisynaptic loop: DG-CA3. Front Neural Circuits. 2013;7:165.
54. Hanuschkin A, Ganguli S, Hahnloser RH. A Hebbian learning rule gives rise to mirror neurons and links them to control theoretic inverse models. Front Neural Circuits. 2013;7:106.
55. Skocik M, Kozhevnikov A. Real-time system for studies of the effects of acoustic feedback on animal vocalizations. Front Neural Circuits. 2012;6:111.

56. Molkov YI, et al. Control of breathing by interacting pontine and pulmonary feedback loops. Front Neural Circuits. 2013;7:16.
57. Chapin JK, et al. Real-time control of a robot arm using simultaneously recorded neurons in the motor cortex. Nat Neurosci. 1999;2(7):664–70.
58. Perel S, et al. Single-unit activity, threshold crossings, and local field potentials in motor cortex differentially encode reach kinematics. J Neurophysiol. 2015;114(3):1500–12.
59. Taylor DM, Tillery SI, Schwartz AB. Direct cortical control of 3D neuroprosthetic devices. Science. 2002;296(5574):1829–32.
60. Wessberg J, et al. Real-time prediction of hand trajectory by ensembles of cortical neurons in primates. Nature. 2000;408(6810):361–5.
61. Kennedy PR, et al. Direct control of a computer from the human central nervous system. IEEE Trans Rehabil Eng. 2000;8(2):198–202.
62. Levine SP, et al. Identification of electrocorticogram patterns as the basis for a direct brain interface. J Clin Neurophysiol. 1999;16(5):439–47.
63. Fernandez-Vargas J, et al. Assisted closed-loop optimization of SSVEP-BCI efficiency. Front Neural Circuits. 2013;7:27.
64. Chaudhary U, Xia B, Silvoni S, Cohen LG, Birbaumer N. Brain–computer interface–based communication in the completely locked-in state. PLoS Biol. 2017;15(1):e1002593. https://doi.org/10.1371/journal.pbio.1002593.
65. Pandarinath C, et al. High performance communication by people with paralysis using an intracortical brain-computer interface. Elife. 2017; 6. pii: e18554.
66. Chaudhary U, Birbaumer N, Curado MR. Brain-machine interface (BMI) in paralysis. Ann Phys Rehabil Med. 2015;58(1):9–13.
67. Jacobs J, Miller J, Lee SA, Coffey T, Watrous AJ, Sperling MR, Sharan A, Worrell G, Berry B, Lega B, Jobst BC, Davis K, Gross RE, Sheth SA, Ezzyat Y, Das SR, Stein J, Gorniak R, Kahana MJ, Rizzuto DS. Direct electrical stimulation of the human entorhinal region and hippocampus impairs memory. Neuron. 2016;92(5):983–90.
68. Hampson RE, Song D, Robinson BS, Fetterhoff D, Dakos AS, Roeder BM, She X, Wicks RT, Witcher MR, Couture DE, Laxton AW, Munger-Clary H, Popli G, Sollman MJ, Whitlow CT, Marmarelis VZ, Berger TW, Deadwyler SA. Developing a hippocampal neural prosthetic to facilitate human memory encoding and recall. J Neural Eng. 2018;15(3):036014.
69. Suthana N, Fried I. Deep brain stimulation for enhancement of learning and memory. NeuroImage. 2014;85:996–1002.
70. Lee VK, Dai G. Printing of Three-Dimensional Tissue Analogs for Regenerative Medicine. Ann Biomed Eng. 2017;45(1):115–31.
71. Zhang YS, et al. 3D Bioprinting for Tissue and Organ Fabrication. Ann Biomed Eng. 2017;45(1):148–63.
72. Hollister SJ. Porous scaffold design for tissue engineering. Nat Mater. 2005;4(7):518–24.
73. Mesko, B. @berci. The ultimate list of what we can 3D print in medicine and healthcare! – The Medical Futurist 2017.
74. @harvard. The Harvard Gazette Creating 3-D tissue and its potential for regeneration. 2016.
75. Tarafder S, Bose S. Polycaprolactone-coated 3D printed tricalcium phosphate scaffolds for bone tissue engineering: in vitro alendronate release behavior and local delivery effect on in vivo osteogenesis. ACS Appl Mater Interfaces. 2014;6(13):9955–65.
76. Tarafder S, et al. Microwave-sintered 3D printed tricalcium phosphate scaffolds for bone tissue engineering. J Tissue Eng Regen Med. 2013;7(8):631–41.
77. Crawford M. Creating valve tissue using 3-D bioprinting. ASME 2013.
78. Lee H, et al. Cell(MC3T3-E1)-printed poly(−caprolactone)/alginate hybrid scaffolds for tissue regeneration. Macromol Rapid Commun. 2013;34(2):142–9.
79. Scott C. Wake forest institute for regenerative medicine progresses with 3D printed skin technology. 3Dprint.com. 2017.
80. Scott C. Chinese researchers believe they can make 3D printed skin a reality in two to three years. Aug 5, 2016|3D Printing.com. 2016.

81. Clegg A, et al. Frailty in elderly people. Lancet. 2013;381(9868):752–62.
82. Cevenini E, Monti D, Franceschi C. Inflamm-ageing. Curr Opin Clin Nutr Metab Care. 2013;16(1):14–20.
83. López-Otín C, et al. The hallmarks of aging. Cell. 2013;153(6):1194–217.
84. GenAge: the ageing gene database. 2018. Available from: http://genomics.senescence.info/genes/.
85. Pan F, et al. Gene Aging Nexus: a web database and data mining platform for microarray data on aging. Nucleic Acids Res. 2007;35(Database issue):D756–9.
86. gene db. 2018. Available from: http://www.uwaging.org/genesdb/.
87. Burkle A, et al. MARK-AGE biomarkers of ageing. Mech Ageing Dev. 2015;151:2–12.

Chapter 8
Business Model in Digital Medicine

Jean-David Zeitoun and Anne Osdoit

8.1 Introduction

8.1.1 Conventional Business Models in Health Care

Business models of conventional health-care products (drugs, medical devices, software) are well established and have been extensively studied. Three main categories can be described: coverage through insurance, out-of-pocket from consumers (i.e., patients or their caregivers), and direct selling to health-care professionals or organizations (mainly hospitals).

The first option is complete or sufficient coverage through insurance, either public and (or) private insurance. This is the case for the majority of drugs and for many medical devices, and/or interventions, in particular so-called high-risk or class III medical devices (according to the US Food and Drug Administration classification, e.g., pacemakers, cancer diagnostics, insulin pump). Insurance, regardless of the sector (housing, transportation, finance, etc.), is associated with a perverse effect called moral hazard. Indeed, people tend to behave more dangerously when they know that a third party will pay in case of an adverse event. This is also the case in health insurance and implies that all health-care insurers try to implement a certain degree of regulation and control upon the products and services they choose to cover. For health-care companies, insurance coverage generates both a hurdle and a guarantee. The hurdle is represented by the validation process imposed by insurers for inclusion of their product in the coverage portfolio. Most of the time, it means that health-care

J.-D. Zeitoun (✉)
Inato, Paris, France

Centre d'Epidémiologie Clinique, Hôtel Dieu Hospital, APHP, Paris, France
e-mail: jeandavid@inato.com

A. Osdoit
MD Start, Paris, France

© Springer Nature Switzerland AG 2019

A. André (ed.), *Digital Medicine*, Health Informatics,
https://doi.org/10.1007/978-3-319-98216-8_8

products should undergo a clinical validation process intended to demonstrate that their products are both safe and effective. More recently, a novel discipline called Health Technology Assessment has emerged, embodied in numerous countries by public or private entities. Health Technology Assessors are now responsible for the evaluation of health-care products regarding both their comparative effects and their cost-effectiveness. This is another level of validation that many health-care companies need to go through for obtaining insurance coverage at a wished price. However, once reimbursement has been accepted, access to market is eased for health-care companies since patients and practitioners are somewhat insensitive to prices.

Second, the out-of-pocket model applies to several kinds of drugs and devices. In theory, out-of-pocket spending should be reserved for health-care products for which clinical benefit is low or not sufficiently established to allow a claim for coverage. In particular, so-called over-the-counter (OTC) drugs (meaning that no prescription is needed, as opposed to so-called Rx drugs) are frequently uncovered by insurers. Many primary care drugs are in the OTC field and address fever, dizziness, constipation, etc. In practice however, the out-of-pocket model can be observed if the manufacturer decides not to wait for the too lengthy evaluation process that precedes reimbursement or chooses (whether reasonably or not) to never submit an application for coverage. It occurs when the cost–benefit ratio is judged unfavorable, i.e., if the likeliness of obtaining a reimbursement is thought to be low and (or) if the cost of the process is too high to justify such engagement.

Third, direct selling to health-care practitioners or organizations (either public or private hospitals) is an option for many medical devices, products used in hospitals even if they are not labeled as medical devices, and software. Several reasons may bring physicians or hospitals to purchase such products:

- Costs can be passed on to patients through an increase in medical fees.
- Payment system of hospitals can be leveraged—or sometimes trumped—to increase revenues through billing modification.
- The newly acquired product can increase efficiency, thereby compensating its acquisition cost: it can shorten hospitalizations and reduce costs, and possibly allow physicians and hospitals to increase their volume of activity.
- Last, a new device can be purchased for marketing purpose. For instance, surgical robots have been incorporated in many urologic teams despite an initial low level of evidence in the demonstration of strong clinical benefit. However, real-life observation suggests that patients now seek robot-assisted surgery, in particular for prostate interventions.

8.2 Digitization of the Economy

It is obvious that digital products and the digital industry have had a huge impact on production, distribution, selling, and use of goods and services in almost all industries, even if not all sectors have been affected in similar degrees. Digitization of the economy has been associated with several major trends, such as the preference for use rather than owning, or the emergence of data as the new source of value, both for prediction and targeting, and for optimization. Some players have been negatively altered by digitization such as media, while others still resist through self-transformation. Also, digital companies are investing in development of goods (cars, devices, physical stores for e-commerce companies, etc.) and conversely, traditional companies are implementing digital services intended to be combined with their historical products.

This so-called fourth industrial revolution has been associated with changes in business models, even if current standards may still evolve in the future. The health-care industry is probably quite immature as compared to other sectors regarding its level of digitization. However, the field of digital medicine is being subject of an intense effort of research and investment. According to several consistent reports, investments both from strategic investors such as established companies and from venture capitalists (VCs) are increasing. Estimates vary from approximately $4 billion to $6 billion every year, depending on the selected criteria. Although promising, the business models of digital medicine companies are not established, and competition for capturing value is a challenge as governments and payers struggle to contain budgets and therefore exhibit a low willingness to pay for an additional set of products and services.

8.2.1 Potential Business Models for Digital Medicine

Business models for digital medicine products might not be so different from other conventional medical products. However two comments can already be made at this stage. First, it seems virtually impossible to accurately predict in which proportions the three main above mentioned models (insurance coverage, out-of-pocket, and selling to health-care people or entities) will be allocated. Second, other novel models might emerge, in particular through so-called indirect selling, namely free access to a digital medical service and subsequent monetization of collected data to another stakeholder.

8.2.1.1 Insurance Coverage

So far, few digital medicine companies have gained such status for their product, but their number is likely to grow as many promising companies seem to engage in a rigorous clinical validation process. WellDoc (St. Paul, MN, USA) developed BlueStar, a digital medicine product targeting people affected by type 2 diabetes, which can be reimbursed by several insurance companies based upon prescription. Omada Health (San Francisco), another pioneering company in the field, has also established partnership with several major insurance companies in the USA, such as Kaiser Permanente (Oakland, CA, USA) or Humana (Louisville, KY, USA).

Engaging in such an avenue has similar pitfalls and advantages compared with traditional medical products. If they want to be able to claim for coverage, novel companies need to develop their products in a way that demonstrates a favorable risk–benefit balance. In summary, it means that they must conduct appropriately designed clinical trials both showing safety and efficacy with respect to relevant endpoints. This process is well defined for drugs and medical devices, and the most ambitious digital medicine products should inexorably adopt similar development approaches. This implies costs that are orders of magnitude higher than those usually incurred for non-medical digital products, yet empirical evidence suggests that the amount of money needed for proof of efficacy is relatively low on average as compared to traditional medical products (either from pharmaceutical or medical device companies). In addition, and as mentioned above, if they plan to compete with other standards of care or even if they seek to be incorporated in the current armamentarium as an add-on, digital medicines need to generate two more types of data. First, they must show a comparative advantage, either as stand-alone products or in addition to the standard of care. This may focus on efficacy, safety, or even convenience for patients (otherwise called practicability in medicine). Second, it seems obvious that economic stakes cannot be ignored. Ideally, digital medicine should be able to generate savings for insurers, either in reducing the need for costly medical visits or unnecessary care, or through outcome improvement and avoidance of costly hospitalization and (or) complication. In this best-case scenario, we anticipate that digital medicines will be able to capture a lot of the value created and to become rapidly highly profitable. Another option and probably the minimum that digital medicines must show is cost-effectiveness rather than being cost-saving. Put another way, it means that the additional cost incurred by third-party payers is worth the value that is created in terms of health outcome as compared to other existing options. In any case, the pricing strategy is paramount and, again, empirical evidence is too scarce to allow an easy and reliable prediction. On the one hand, and as for other digital companies, marginal production costs are low. On the other hand, we believe that companies developing digital medicines will target an ambitious pricing strategy—possibly disconnected from their own costs—so as to capture as much value as possible, especially if their products are likely to generate savings for third-party payers.

Two major trends related to coverage should be underlined. First, it is undisputed that chronic diseases and (or) conditions linked to aging already account for the majority of the medical and economic burden in advanced countries. Most health-care systems are ill-equipped to face their increasing importance, and disruptive solutions will be needed in the coming years to overcome this challenge. Digital medicine products are obviously well positioned to reach the needed objectives, and insurers should be willing to invest into projects that are likely to sustainably divert patients from costly care episodes such as hospitalizations or interventional procedures. Second, and consequently, many health-care systems are trying to shift their preeminent model from a fee-for-service to a fee-for-outcome (or fee-for-value) scheme. Bundled payment for surgical procedures is an illustration of such evolution. Payers are more and more interested in incentivizing providers or health-care organizations toward cost-effective management. This raises at least two kinds of challenge.

First and more obviously, it implies that physicians and hospitals implement procedures likely to deliver a better outcome despite restricted expenses. Second and perhaps even more technically difficult, it means that novel and specific capabilities must be developed in order to reliably measure and analyze a great amount of data likely to demonstrate the value delivered. Digital technologies are thought to be an evident means of gathering such data, thereby making digital medicine products in an ideal position to be aligned with the emerging fee-for-outcome model.

8.2.1.2 Out-Of-Pocket

According to a nationwide survey, Americans spent approximately $30.2 billion out-of-pocket on complementary health approaches. However, this represents less than 10% of all out-of-pocket spending, and in the health-care field, patients have historically been one step removed from payment, even though there are significant differences between countries. Yet many digital medicines already fall under that status. According to a report of 2016 from the Silicon Valley Bank, 55% of digital health investments since 2011 have been in companies whose products interact with the consumer, even if it does not necessarily imply that this latter is also the payer. One partial explanation might be similar to the one described above for other conventional medical products, namely, a low likeliness to demonstrate sufficient value or a lack of willingness to engage into a lengthy and costly process for clinical validation needed for reimbursement. Another possible reason might be related to historical standards in the digital industry. Indeed, most digital markets are loosely regulated and digital entrepreneurs are not used to regulatory processes. Consequently, some might expect that patients will behave as regular consumers and therefore purchase a digital product if they perceive any kind of personal advantage in doing so. Even if this model is likely to perpetuate, entrepreneurs and their investors should be aware that patients and their caregivers might not increase personal spending in a limitless way. According to a large survey focused on digital health and conducted by Rock Health (San Francisco, CA, USA), while 52% of people agreed that they are responsible for their own health, only 7% said they would be willing to pay out-of-pocket for health care. Digital medicine companies aiming to construct a business model based upon out-of-pocket payment should be aware that securing premium pricing will be difficult and that going through the regulatory process might not be an option. In fact, it seems unlikely that such companies will be able to gain significant market uptake if they address a major medical condition for which health care is generally already covered.

8.2.1.3 Direct Selling to Health-Care Professionals or Organizations

This model is likely to increase if digital products obtain indirect coverage through payment to physicians or if they create value in another manner from the medical entity standpoint. The first option, i.e., granting of an "indirect" coverage, seems to be an emerging way for digital medicine products to gain market traction. In several countries, insurance schemes specifically designed for patients with chronic conditions cover a predefined amount of annual spending for remote solutions aiming to increase adherence and, more generally, quality of care. In France, for instance, public health insurance (which is a state monopoly) is experimenting payment to physicians for what they call "telesurveillance" services for patients with connected pacemakers. A predefined amount of money will be allocated to physicians for each patient monitored on a yearly basis. It is then incumbent to the physician to subscribe a telesurveillance solution with an appropriate company. Even though everything is not established, fees that will be dedicated to physicians should allow profitability both for digital medicine companies and for health-care professionals so as to align incentives for every stakeholder. Implicity, a Paris-based company that develops an artificial intelligence solution reducing false alarms issued by pacemakers, has started to deploy over many hospitals, thanks to that novel payment pathway.

The second option for digital medicine companies is to generate value for health-care entities (either people or organizations such as hospitals) if they seek to sell them their products. We consider three avenues of doing so. First, digital medicine products can, as already said, increase quality of care (*clinical value*). From that standpoint and since health-care practitioners pursue their endeavor toward health improvement for both intrinsic and extrinsic reasons (e.g., reputation), they may want to pay for digital solutions which have demonstrated in a compelling manner a clinical benefit, in terms of either efficacy, safety, or practicability for patients. However, the pricing of such solutions will be a critical issue since, on the other hand, health-care professionals or their hospitals will certainly scrutinize their financial balance. Second, digital medicine products can ease organization and work flow for physicians and the clinical units they practice in (*organizational value*). Benefits of such improvements are manifold: enhancement of patient experience within care episodes; improvement of quality of life at work, which is likely to increase attractiveness of equipped centers; and increased productivity. Third, digital medicine products may in such cases help physicians or hospitals to generate higher revenues (*financial value*). Besides the abovementioned productivity gains, digital products can allow physicians to manage patients with conditions for which coding brings higher payment, to charge greater fees if perceived clinical benefit is aligned, or to mutualize several medical interventions such as in rehabilitation centres.

8.2.1.4 Partnership with Established Companies

Established companies such as pharmaceutical companies and medical device companies have a great deal of expertise in several fields that are useful for younger digital medicine companies: clinical development, regulatory affairs, market access, selling, communication, and even post-marketing product surveillance. Those skills are undoubtedly critical and are a strong factor driving some partnerships between small companies and big companies. Small companies see in their established counterparts a source of acceleration and enhancement of success likeliness. Another non-mutually exclusive advantage is to raise funds through Corporate Venture Capital, whose number at the global scale is on the rise. Last, it is also a way to acquire visibility and gain access to final users, either patients or physicians. For established pharmaceutical companies or medical device companies, partnering with a small company is a means to generate growth for themselves in the mid- or long term. Investing in external innovation and combining it with internal innovation is the theoretical model, assumed to increase the offer of products and services of the older company. External investments are supposed to offset the lack of self-originated innovation and also to prepare big companies to emerging types of markets that go beyond their traditional activities. Voluntis, a French company that was founded in the early 2000s and that has recently gone public, develops so-called digital therapeutics to accompany and enhance drugs or devices, in particular in the fields of diabetes and oncology. Their software or and smartphone applications allow patients to send to their physicians some information related to biological parameters or treatment intake, and to receive medical advice intended to optimize disease management. Voluntis has established partnerships with several pharmaceutical companies (Sanofi, Roche, Astra-Zeneca), which has been seen by analysts as an advantage over some competitors before its initial public offering.

8.2.1.5 Monetization of Data

Currently, high-quality data collection at a large scale is a very expensive process, being unaffordable for many entities. It drives rising costs of clinical trials. On the other hand, novel digital technologies are able to collect large-scale data with minimal investment, yet the quality, preciseness, and validity of such data are low on average or at least uncertain. Crossing that chasm is plausible in the near future and likely to allow clinical studies to be conducted at more reasonable costs, either for marketing approval or designed after commercialization. Also, gathering real-life data in a massive way about patients' diseases and their routine management is likely to be of interest for some big companies such as pharmaceutical companies. It might help them to increase their knowledge and understanding of patients, conditions, and treatments, thereby allowing them to properly identify target patient populations, to tailor their marketing approaches or to gain insights useful for R&D. For instance, by better knowing mechanisms of early disease onset, they could better design drugs to be incorporated in their clinical development programs and could also design their clinical trials with a methodology more likely to demonstrate the intended effect. Arivale, a Seattle-based company, is one example of such an approach. Individual's health is measured through four key features: DNA, biological markers (blood and saliva), gut microbiota, and lifestyle data. At the beginning, its business model relied upon direct payment on an annual basis from patients seeking to augment their knowledge and eventually wellness. Then, it seems that the company diversified sources of revenues through partnership with medical products companies or employers to respectively conduct clinical studies or improve employees' wellness.

8.3 Conclusions

Digital medicine, and in particular its different potential business models, is immature and therefore remains a rapidly evolving field. Technological advances often go faster than their use and even more quickly than regulatory reforms. Whereas there has already been a great amount of investment and progress over the last decade, still little is known about optimal current and future ways of profitability and reasonable value-sharing. However, it seems that after an initial wave of reject of traditional business models inherent to the health-care industry—in particular insurance coverage—digital entrepreneurs in the field are more and more seeking to gain not only regulatory approval but also reimbursement by usual payers. As such, their products often follow the well-known pathway of clinical development and value demonstration. In order to be widely adopted, those digital medicine products, that are considered as medical devices for many of them, need to show that they are first safe and effective, and second that they enhance health outcome and ressource utilization. Besides technological progress, two main factors drive this trend: first, some limitations of current medicines (drugs, devices, interventions, etc.) are to be overcome through a better use and organization of existing tools, or through novel tools. Digital medicines are enablers of such optimization. Second, increases in health-care costs are an enormous concern for governments. Many drug prices are skyrocketing, some devices are also becoming more and more expensive, and foremost, demographic trends predict a growing burden of diseases. With health-care systems unfit for many population health purposes, digital medicine products are seen as ideal tools for fixing some features of the system. However, challenges are huge since health-care systems will still need to deeply change if they are to realize the full potential of many coming digital advances. Whether they will steer themselves toward the needed evolution, or pass it on to health-care organizations and practitioners through mandatory productivity gains remains to be determined. Last, novel business models are also to be explored since the digital industry works according to different rules than more conventional hardware-based industries. This must be taken into account. Potential for both efficiency gains and outcome improvement is huge through optimization and a better targeting of health-care interventions. Any stakeholder must pursue its effort to design novel types of value-sharing models that would both reward innovators as expected and deliver either individual or collective clinical benefits.

Chapter 9
New Technologies, Telemedicine, eHealth, Data...What Are You Talking About? The Lawyer's Point of View

Lina Williatte-Pellitteri

Today, it is undeniable that new information and communication technologies have been widely introduced in hospitals and clinics as well as in medical practices. The doctor, who is then connected, is vigorously accompanied on a daily basis by his patient, sometimes asking for a 24-hour medical surveillance.

This digitization of doctor-patient relationship, however, developed in different sequences across the world. Leadership belongs to the Anglo-Saxon countries and especially the United States, today in the run for the realization of a big data health care, while in Europe and particularly in France, one wonders about the consequences of a possible establishment of a highly regulated open data health.

The difference of the approaches is poorly understood by doctors wishing to develop scientific research projects like their Anglo-Saxon colleagues who feel constrained by an opaque norm; the patients, amateurs of quantified self-coaching and remote medical coaching, or industry, are forced to consider the place of development of the project according to the degree of normative constraint imposed by the states.

Actors are legitimately denouncing fundamentally different standards, in that they confer, according to their libertarian nature or not, competitive advantages, not catch-up. Nevertheless, the lawyer is able to explain these various normative conceptions according to the legislative traditions of the countries and their policies to protect the fundamental rights of the traditions which are fundamentally divergent, thus explaining the applicable normative framework and developments in medical practices and services through new technologies on the one hand and the processing and use of data that emerges from it on the other.

L. Williatte-Pellitteri (✉)
Lille University, French Telehealth Society, Paris, France

© Springer Nature Switzerland AG 2019

A. André (ed.), *Digital Medicine*, Health Informatics,
https://doi.org/10.1007/978-3-319-98216-8_9

9.1 Telemedicine Versus eHealth: The French Particularity

Telehealth: A Word of Various Senses

Nowadays, one cannot deny that terms, expressions, and words are used indifferently and without distinction of meaning to qualify acts, trends, and practices that are generally related to telehealth. As evidenced, a recent study [1] revealed that the terms telemedicine, telehealth, and eHealth are used as synonyms, [2] while in some cases, they designate a remote medical practice and define the act of care [3] or refer to the tools and therefore the intermediation view of service [4] production or health-related e-commerce [5]. Different meanings words that are not without impact on the applicable legal framework and on the perception of than the legislator of the said act. To illustrate, let's take as an example the typical case of telemedicine. Two conceptions of telemedicine are classically opposed: in developed countries, telemedicine is essentially clinical, whereas in developing countries, it is more informative [6]. Clinical telemedicine is understood as "a professional activity which implements means of digital telecommunication allowing doctors and other members of the body to remotely perform medical procedures for patients," [7] while informative telemedicine is defined as "an audiovisual communication service interactive platform that organizes the dissemination of medical knowledge and protocols for the management of patients and care in order to support and improve medical activity." [8] France is one of the first countries to have a legal basis for the practice of telemedicine through Law No. 2009–879 of July 21, 2009, known as the HPST law, and its October 19, 2010, decree [9]. These two fundamental texts devote a clinical approach to telemedicine, defining it as a medical act performed by health professionals via information and communication technologies. In other words, telemedicine in France is a medical procedure performed by a health professional; its efficiency is recognized and it guarantees safe and proven medical knowledge. This definition traces a clear boundary of how other countries refer to telemedicine as any "health benefit" achieved through new technologies. The word difference, although minimal, is fundamental in that one refers to it as a medical act and the other a medical service. The word difference results in an applicable regulatory framework in particular. Therefore, in France, telemedicine comes under a specific legislation, whereas in Europe, it falls under the more general category of eHealth [10].

9.1.1 Telemedicine: A Medical Act Carried Out Via IT

Telemedicine: A Medical Act

The French conception of telemedicine makes it a medical procedure carried out by health professionals via information technologies. The following is a telemedicine principle confirmed by Article R.6316–1 CSP (Public Health Code): "Medical procedures performed remotely by means of a device using the information and communication technologies."

The 2010 decree lists different telemedicine acts:

- Teleconsultation: This allows the medical professional to provide consultation to a patient from a distance.
- Tele-expertise: This allows the medical professional to remotely solicit the opinion of other medical professionals (...) on the basis of medical information related to patient care.
- Medical telemonitoring: This allows the medical professional to interpret remote data necessary for the medical follow-up of a patient and, where appropriate, make decisions about how to take care of this patient.
- Tele-medical assistance: This allows the medical professional to attend remotely to another health professional during the medical act.
- Medical regulation: This is the answer of the doctor brought within the framework of the medical regulation (related to French "SAMU," emergency medical service with medicalized ambulances).

In addition, for the telemedicine project to be acceptable, it must take into account the deficiencies in the supply of care due to the insularity and geographical isolation of the territory (Article L6316–1 CSP). Its activity and organization must be the subject of either a national program or a local contract [11] (CPOM, CAQCS) concluded in particular with the regional health authority [12] (ARS).

Thus, in France, telemedicine was earlier thought by the legislator as a "tool," in particular, to meet the objectives of organization—continuity, permanence, and security care—but also basically as a medical act, wherein the conditions applied in medical practice must be planned and respected. As such, the 2010 decree project holders must respect the fundamental rights of the patient and the obligation of health professionals to meet the legal requirements for medical practice provided for by the Code of Public Health and the Code of Medical Ethics. The set is valued and verified by the authorities, if appropriate responsibilities [13] may result. These regulatory constraints are forging an identity of their own and of telemedicine which differs from eHealth benefits.

9.1.2 *eHealth: A Service Provided Through the New Technologies of Information and Communication (NICT)*

eHealth is defined as a health service delivery linking a service provider and a consumer. More generally, the majority of these are personal services related to the information society including e-commerce. The field of the application is vast: tele-observance of patients treated at home by health providers, tele-medical advice, availability of health information on the Internet, and health or well-being coaching (Mobile Health). These benefits conceived as a service and not as an act in France are excluded from the scope of the 2010 decree and the European Directive of June 8, 2000, the so-called e-commerce directive, which was transposed in France by the law of June 21, 2004, on digital trust [14]. Consequently, the normative constraints respond less to the imperatives of organization, health care, and security, like the texts applicable to the practices of telemedicine, that of free movement, ease of realization, and remuneration of benefits.

All of this justifies that eHealth projects are easier to realize because of few constraints from a regulatory point of view. Does this mean that the conception adopted by the state in its practice of telemedicine may depend on its development and economic attractiveness? The answer is without a positive reservation, except to add that from the constraint to the project. This results in the quality of the practice and the service rendered to the patient. The example of medical tele-consulting is demonstrative on this point. If in France tele-medical advice is not part of the acts provided for by the 2010 decree, and is not a qualifiable telemedicine act, it is because its design does not allow protection of the person in demand, on the one hand, and, the professional solicited, on the other. Thought of as a commercial service, tele-consulting is therefore part of eHealth and intends to be less protective of the rights of the patient than of the rights of the consumer. The reciprocal is true also for the health professional who is therefore considered to be a service provider subject to the constraints of e-commerce. Thus the applicable standard is dependent on the protection of the actors and beneficiaries of the medical activities and services carried out via the ITs. The normative constraint then becomes a strategic tool for development. The example of data processing resulting from these new practices also substantiates.

9.2 The Regulatory Framework Applicable to the Data

Health TV and its various forms of realization, insofar as it supposes the use of information and communication technology, inherently generate data, which contains raw information revealing an activity or a certain state at a specific time.

One cannot deny that in recent years, there has been an acceleration of production of both national and supranational law relating to data exploitation; this verdict is revealing the stakes that today represent digital data processing. That these issues are economic, financial, societal, and legal, the data is, unquestionably, at the heart of the day-to-day concerns of entrepreneurs. Well named, because far from being only raw information, its treatment can be the source of a serious violation of the fundamental rights of individuals. For this reason, a normative framework is essential.

In France, it is the computer and freedom law of January 6, 1978, [15] which laid down the fundamental principles applicable to personal data processing. This law has been first amended by the law of August 6, 2004 [16], the main purpose of which was to transpose in France the European Directive of October 24, 1995 [17]. This same directive followed a recommendation of the Council of Europe which had clearly stated the need to harmonize the data processing provisions of the Member States in order to limit the obstacles to their free movement within the European Union. Despite these texts, different practices have been observed in European countries, with certain states having proved more permissive than others, and development facilities are denounced by actors constrained by a stricter national standard in the matter. It was not until the Charter of Fundamental Rights of the European Union that the right to personal data protection in European law was granted [18]. Adopted by the Nice Intergovernmental Conference, the Treaty of Lisbon rendered it binding [19]. The set was consecrated by the European Regulation of April 27, 2016, relating to the protection of individuals with regard to the processing of data of a staff and to the free movement of such data [20]. The legal value of a European regulation is more binding than that of a European Directive. Indeed, if a directive sets a flexible framework allowing Member States, which committed to apply it, to draw inspiration from it to legislate on their territory, the European Regulation would be more opposable. Therefore, the margin of interpretation and state appreciation is much narrower so that the national law to be adopted in the application of the said Regulation either completes it or specifies it but in no way contradicts it. This is the essence of the text of the law for the Digital Republic of 2016. Therefore, if it can be said that the French and European legal frameworks now converge on the foundation of personal data protection, then it should be emphasized that peculiarities distinctive to the French state remain.

9.2.1 General Framework

It must be remembered beforehand that unlike the United States, no property right has been recognized on the data in Europe and therefore in France. In fact, the data is not legally described as a "thing or object"; therefore, it does not benefit from the prerogatives granted usually to the owners. Thereby, all possibilities of claiming a property right over the data and thus any right of purchase or sale are excluded.

According to the European vision, data is an emanation of the legal personality of a physical person. It falls under the category of subjective rights. Like the identity of a person, it enjoys the same legal protection as the information privacy of people. The right to the protection of the privacy of individuals is one of the most protected rights of our normative system. Dedicated by Article 12 of the Universal Declaration of Human Rights and taken up in Article 8 of the European Convention on Human Rights in France, it is contained in Article 9 of the Civil Code which was recognized in 1995 as a constitutional value [21].

Therefore, any harvest, treatment, exploitation, transfer, circulation, etc., of a given personal information is subject to the prior information and consent of the person it is from.

9.2.1.1 Qualification of the Data

The application of this principle therefore implies that the data is a personal data relating to the identity of a person. Anonymous data does not require, for its treatment, the consent of anyone from whom it emanates. However, it is still necessary to be able to qualify the "anonymous data." An affirmation does not seem to be easy today, as the performance of IT solutions and tools undermine any hypothesis absolute anonymization. Therefore, lawyers recommend to make information and consent of the person a principle applicable to any treatment, even if it does not only concern so-called anonymized data.

In any case, the burden of proof of consent rests on the person responsible for the treatment. The request for consent must be presented in a way that it is distinguished from other questions that the controller may pose and should be formulated in an understandable form, accessible, and in clear and simple terms. Moreover, the consent thus collected must have been given freely, that is, without constraints, without manipulations, without false information like the service provider who packages and supplies a service or the performance of a contract with the consent of the person to the treatment of his data, even though the said processing is not necessary for the performance of the service sold or in the performance of the contract.

9.2.1.2 Special Cases

Holding the Data Subject

This is mainly to report the situation of the minor child. The European Regulation did not prohibit the processing of the data concerning them. Nevertheless, some limits to secure it were set. Thus, the processing of a minor's personal data is lawful when he is at least 16 years old. Below, the consent of the holders of parental authority is required. It is the responsibility of the controller to provide proof of compliance with these conditions.

Holding to the Qualification of the Data

The processing of personal data is authorized, provided that the informed person agrees. The principle is reversed when the processing involves so-called sensitive personal data. Qualified as sensitive are any data that reveals the person's racial or ethnic origin, political opinions, religious or philosophical beliefs, or trade union membership; genetic and biometric data, which are used to identify a person in a unique way; health or life data; and data that reveals a person's sexual orientation.

The processing of so-called sensitive data is prohibited. This principle is accompanied by exceptions provided for in Paragraph 9 of the General Data Protection Regulation (GDPR) [22], including the explicit consent of the data subject for the purposes authorized by the European Union. However, the National Commission on Computing and Freedom (CNIL), the supervisory authority, must first, at the launch of the treatment, allow treatment. In other words, the consent of the person is not a sufficient condition in itself; for sensitive data processing, it is imperative to file an application for authorization with the CNIL, which will only grant it in particular, a guarantee of protection of the person's private life and legitimacy of treatment and its purpose.

9.2.2 *Rights of the Data Subject: Basic Principles of Data Processing Personal*

Data is part of a person's private life. Therefore, the rights attached to it have the same protection. This protection implies the respect of a certain number of requirements that the controller must guarantee. Thus, the person whose data is processed must be informed on the following:

- The identity and contact details of the data controller, his representative, or the data protection officer (DPO [23]) where applicable.
- The right of access to processed data, the right of rectification, the right to erasure, the right to object, the right to limitation of treatment, and the right to portability of data.
- Purposes of the processing, the data retention period, and the existence of automated decision making, including profiling.
- Recipients of the data and if a data transfer to a foreign country (Europe or outside Europe) will be done.
- The right to lodge a complaint with a supervisory authority.

- *A right to information for...*

The person who consents to the processing of his/her data must be informed of the purposes of the said processing in order to issue an informed consent. Although apparently easy to achieve, this requirement is not, in the field, easily held. In fact, the controllers and/or the project leaders do not think sufficiently of the potential of their project that at the moment of its deployment, they use the data for a purpose other than what has been granted by the person concerned. It is good that analyzing this behavior as a failure is legally justifiable, and it is likely that responsibility will be put on the controller.

In addition, this right to information also concerns the right of every person to know the final destination of his/her data, especially if he/she is going to cross the borders of France, Europe, or beyond.

The transfer of data is envisaged either outside the French borders or within the European Union. The obligation of the European states to comply with the GDPR in May 2018 at the latest is likely to secure the data protection of the dice when such data are processed in a country that has committed to the same level of protection as that of France. To this end, a European Data Protection Board is established primarily to ensure the consistent application of the regulation within the Union.

The transfer of data outside the borders of Europe is not possible until the European Commission has determined by decision that the third country ensures an adequate level of protection. In the absence of a decision of conformity of the Commission, the controller or the processor cannot transfer the data to a third country unless it has provided appropriate safeguards for the protection of the rights of individuals. These guarantees can be legally binding and enforceable by authorities, binding company rules, standard data protection clauses adopted by the Commission or by a national supervisory authority approved by the European Commission, a code of conduct, or a certification mechanism.

- *A right of access to the data for...*

At any time, a person has the right to obtain from the controller any information which allows him to assert his rights. In this case, it is a matter of verifying the transparent character and loyal treatment. This is how the person is entitled to obtain responsibility for processing the correction of the data relating to it, as soon as they prove to be inaccurate. She can also get the limitation of treatment when it disputes the accuracy of the data concerning it or when it identifies an unlawful treatment. It may also require deletion of data either because it is not necessary in view of the purpose of the treatment or because the person has withdrawn consent and/or opposes the processing of the data, for the reason that the processing is found to be unlawful. Nevertheless, the person may be refused the execution of this right when the processing of the data meets the public interest requirements or the exercise of the right to freedom of expression and information [24].

- *A right to portability of data for...*

This is one of the novelties of the GDPR. The right to portability implies that everyone, whose data is processed in a computerized manner, is entitled to retrieve the data it provided to the controller in a structured, commonly used, and machine-readable format. However, this right is only applicable to computerized processing which presupposes upstream the consent of the person (thus excluding treatment by computers in response to a mission of the general interest or public body authority responsible for the processing).

9.2.3 Accountability of the Controller (The Principle of Accountability)

Although previously based on a system of compliance with legal obligations that it belonged to national authorities to monitor (in this case for France: the CNIL) to verify and sanction, where appropriate; the GDPR modifies the postulate and puts in responsibility upstream the project, the controller. From now on, it is up to the latter to implement data processing in conformity to the normative requirements and to detect any failings, to declare them, and to apply corrective measures. Therefore, the controller must proceed with self-diagnosis, on the basis of an impact analysis, of its data processing system both on its security and on its legitimacy. In the event that the treatment is at high risk, the controller must submit measures to the CNIL to reduce it. Otherwise, the CNIL can accompany the controller for compliance, and case penalties may be imposed. At this point, the GDPR has also innovated. Thus, if a controller or a subcontractor violates willfully or negligently obligations incumbent on it, the total amount of the administrative fine may be raised up to EUR 10,000,000 or, in the case of a company, up to 2% of the total global annual turnover of the previous financial year. This penalty can be up to 20 million EUR or for companies up to 4% of annual worldwide turnover for the precedent year where the violation relates to the basic principle of a treatment and in particular the conditions applicable to the consent and the rights of individuals. The severity of the penalty demonstrates the European legislator's resolution to enforce the legal framework for the computerized processing of data—a framework that *ultimately* embodies two new principles:

- The principle *of accountability* for the controller who is now responsible for the lawfulness, legitimacy, and security of data processing, not only vis-à-vis the supervisory authority (the CNIL) but also with regard to the person whose data is being processed.
- The principle of *empowerment* of the people whose data is processed. By dedicating the basis of rights, the legislature thus allowed the person concerned to exercise control over the processing of his data. This control aims to make the person responsible in the management of his data.

9.2.4 The Particularities of French Law

The Law for a Digital Republic of October 7, 2016, made in application of the European Parliament regulations of 2016 has of course included in its subject the content of the GDPR, moreover some aspects, she went further. Thus, it recognizes every citizen's right to free provision of his personal digital data that comes in several concrete measures: the right to digital oblivion for the minor (thus, a minor can more easily and quickly delete online content about themselves) and the right to digital death (every citizen will have the right to express, during his/her lifetime, his/her wishes on the conservation and communication of his/her data after his/her death or request it to be deleted).

Also, the strict application of the private correspondence rule forbids all access by an unauthorized person to the contents of correspondence regardless of its vector or the information technology used (mails, etc.).

It is clear that we are in the yard of the construction of a right of treatment computerized data. A right that will be advised to entrepreneurs to master before the deployment of their project because experience shows that any incident following the non-respect of the normative framework applicable to the processing of the data is at the origin image damage that shifts the company's relationship of trust with its customers in defiance relationship. A rocker that is difficult *to retrieve* afterwards.

References

1. Gallois F, Rauly A. Telemedicine and comparison of health systems: methodological issues. In: Colloquium international research & regulation 2015, 10–12 June 2015, Paris.
2. Crigger BJ, Policy & Politics. e-Medicine: policy to shape the future of health care. Hastings Center Rep. 2006;36:12–3.
3. Chandra A, Pettry CE Jr, Paul DP. Telemedicine from a macromarketing viewpoint: a critical evaluation with proposed licensing strategies. In: Government policy and program impacts on technology development, international trade and transfer outlook; 2013, p. 111–136.

4. Kerleau M, Pelletier-Fleury N. Restructuring of the healthcare system and the dissemination of telemedicine. Eur J Health Econ. 2002;3(3):7.
5. Kirsch G. The business of eHealth. J Med Market Device Diagn Pharmaceut Market. 2002;2(2):106–10.
6. Simon P. Telemedicine is not e-commerce. Communication from ANTEL in partnership with the National Order of Physicians. 22 Nov 2013.
7. Croels JM. The law of obligations to the test of telemedicine. PU of Aix-Marseille, 2006, p 38, cited by Simon P. Telemedicine ce is not e-commerce, *ib.*
8. Simon P. Telemedicine is not e-commerce Ib. p. 39.
9. n ° 2010-1229 JORF n ° 0245 of October 21st, 2010.
10. This is the case for example in Germany: *in* http://www.science-germany.com: e-health in Germany.
11. who can be a CPOM: multi-year contract of objectives and means: signed between the director of a health facility and the Authority Local Health Authority or a CAQCS: Contract for Quality Improvement and Care Coordination concludes between the project sponsor and health insurance.
12. In this case, the Regional Health Agency.
13. Mainly, civil liability to compensate for harm suffered by the patient due to non-compliance with the standard or liability ordinal for the health professional, offender.
14. The purpose of this law has been to create a legal framework to ensure the free movement of information society services.
15. Law No. 78-17.
16. Law No. 2004-801.
17. Directive 95/46 / EC on the protection of personal data.
18. Charter of Fundamental Rights of the European Union 18/12/2000. OJ of the European Communities C 364/1.
19. Treaty on European Union and Treaty on the Functioning of the European Union which adopted the Treaty of Lisbon signed on 13 December 2007 published in the Official Journal n ° C326 of 26/10/2012, p.0001–0390.
20. No. 2016/679. JOUE n ° L119 of May 4th, 2016, p.1. This directly applicable regulation in the Member States' internal legislation repeals the Directive 95/46 / EC as obsolete.
21. Cons. Const. Jan. 18, 1995, No. 95-352 DC. OJ 21 Jan. 1995.
22. GPDR: (European) General Data Protection Regulation.
23. Mandatory as of May 2018.
24. Paragraph 3 of Article 17 of the GPDR specifically provides that the right of erasure does not apply if the processing is necessary. The exercise of the right to freedom of expression and information, (b) to fulfill a legal obligation which requires the treatment provided for in Union law or by the law of the Member State to which the controller is subject, or to perform a public-interest mission. The exercise of public authority of the controller, (c) for reasons of public interest in the field, (d) for archival purposes in the public interest (...), (e) the recognition, exercise, or defense of rights in justice.

Chapter 10
Conclusion: Future Trends for Our Health-Care System

Arthur André

Digital revolution undoubtedly modifies the way we develop, practice, and provide medicine. With this paradigm shift, several parameters will directly influence the evolution of health-care systems:

- The data and the property of the data. Who will manage these new data generated? Who will analyze them and make them intelligible in the medical sense? Is it the citizen himself? Is it the State? Is this a new private actor of the system, having enough technological advance, like the American GAFAMI (Google, Apple, Facebook, Amazon, Microsoft, IBM) or Chinese BATX (Baidu, Alibaba, Tencent, Xiaomi)?
- The transition from curative medicine to preventive medicine. Is there not a risk of medicalization of society? Of eugenics? After curbing pathological functions, will there not be temptation to improve the healthy individual as proposed by transhumanism. How far, technically and ethically, can the sorcerer's apprentices be played with the species?
- The cost of precision medicine. Who will have access? Who will pay? Should we redefine the relevance of care, based on efficiency, which will ensure for a greater number the most adequate care by limiting excesses and medical wanderings?
- The place of the individual in this new system. The global sharing of knowledge and transparency are in line with a health democracy. Should the patient be the main actor, the conductor of his care? Moreover, do the economic or physiological inequalities brought by access to personalized medicine have to redefine a social contract?

A. André (✉)
Pitié-Salpêtrière Hospital, Assistance Publique Hôpitaux de Paris, Paris, France

Sorbonne Université, Paris, France
e-mail: arthur.andre@neurochirurgie.fr

© Springer Nature Switzerland AG 2019

A. André (ed.), *Digital Medicine*, Health Informatics,
https://doi.org/10.1007/978-3-319-98216-8_10

This asks a lot of questions. Most of the medical tools that feed the good or bad fantasies of the fosterers of the prospective are still in the state of prototype or partially conclusive test. On the other hand, it is certain that the race for technological progress will lead to at least some success.

We can therefore imagine possible scenarios depending on the evolution of these medical technologies and their acceptance by the society.

As many other fields of industry and services, health care is deeply impacted by the digital revolution. A pioneer in personalized medicine, Dr. Eric Topol, describes the "Creative Destruction of Medicine" [1]. This reference to the economic theory of Joseph Schumpeter [2] put the innovators and the public demand at the heart of the revolution, to progressively democratize digital health care with reduced costs. This may happen. Governmental initiatives as the Blue button® [3] in the USA or the DMP in France are generalizing electronic health records and the digital access to health-care data for each individual. Those data will grow exponentially with the data collection at different scales (genomics, metabolomics, physiological parameters, body sensors, etc.) until a personal profile of the individual is created, able to accurately predict risk of rare illnesses or susceptibility for common diseases. Projections say that more than one billion individuals will have their genome sequenced in 2025. The GAFAMI in the USA and their equivalent BATX in China are presently the most realistic organizations capable of analyzing health-care data at a very large scale and producing strong artificial intelligence. Therefore they are the only actors to date to be able to provide global health-care solutions, integrating the sum of startups and research's group innovations throughout the world. It is, for example, the goal of Google's Medical Brain team [4] that had already created a tool that could forecast a host of patient outcomes, including how long people may stay in hospitals, their odds of re-admission, and chances they will soon die, from the data of more than 110,000 medical records, including free-text notes [5].

We can then easily imagine that such a tech company could create robots capable of giving—in real time and at a marginal cost—diagnoses (and treatment) on simple pathologies in the beginning, just as surely as an overworked general practitioner who does not have time to train himself on the progress in medicine knowledge or diversity of proposed personalized treatments. One can also imagine that this tech company could provide to individuals an equivalent of the Blue Button® in return for the authorization to use their health data in a multi-level contractual framework saying "The more you leave me free in the use of your data, the more I offer you services" [6]. The control of the data, if it is health data, leads obviously to a certain and scary control of life. The control of our lives through the control of our vital parameters from our genome to our physiological manifestations can be the expression of a new power that must be controlled.

Controlling the birth rate, the mortality rate, but also the capacity of each individual according to their genetic and epigenetic programming will be able to establish a new biopolitical power, already imagined by the French philosopher Michel Foucault [7], much more powerful than the simple disciplinary power, delivered through laws in a state of law or by coercion and violence in a dictatorship. At the extreme it is a technology of control of the submitted bodies.

The previous scenario forgets one central actor of the health-care system: the patient himself. If the patient does not adhere to the offer of care that is offered to him, it will never work. This phenomenon is already observed with the deployment of telemedicine. Residents in areas with poor access to health care remain suspicious of what they consider to be low-cost medicine, and demand from the public authorities "real" consultations with doctors "in the flesh" as the inhabitants in well-equipped areas [8]. For many users of the health system, the doctor–patient relationship is paramount, and in that case, digital solutions will show their limits: whether it is telemedicine or artificial intelligence, patients will have as an intermediary a machine between medical knowledge and them. This computer, as gifted as it is, will have great difficulty to show as much empathy as a human being. Thus, if we can imagine an algorithm using more and more finesse in the human language, or a teleconsultation close to the real one with the help of holograms, reaching the Turing Test, there remains a dimension of the human communication that the digital does not take (still?) into account: nonverbal communication [9, 10]. Indeed, the expression of the face, a hand on the shoulder at the right moment, and the expression of the look are all means of exchange and empathy strongly brought by the human experience and tested by our species.

Moreover, the number of patients to treat will increase. World population is increasing and getting older in developed countries, which will also increase the global burden of diseases and its costs. A fair state is always looking for a balance between funding and health system spending to ensure a decent basic health care to the greatest number of people, by regulating health care through national welfare, non-profit solidarity scheme, and private insurances. This escapes the traditional rules of capitalism because "who benefits" is not necessarily "who pays". Furthermore, to borrow business terms, there is a mismatch between investment returns timeline and long sales cycles in health care: the benefit of an innovation may have significant effect several years after its deployment. And the benefit may not appear directly: if a precision medicine system allows personalized management of a chronic disease with adaptive monitoring and drug delivery, this could be calculated in terms of healthy survival and quality of life for the patient and in terms of hospital re-admission avoiding, complications, and extra cost for the system. To come back to marketing, the life-time value has to exceed the cost of acquisition.

This is not an unbreakable equation. The interconnections of all these fields, digital science, medicine, finance, and public policy, could be the opportunity to reinvent a virtuous health-care system. As we have seen in this book, digital innovations represent a disruption, but they have to meet regulatory and validation principles to actually work. This is merely obvious in medicine: what really benefits patients will be accepted. We must authorize, validate, evaluate, and manage telemedicine, online medical information, diagnostic artificial intelligence, remote monitoring, and robot surgeons. The efficiency of the system should be an equilibrium between value-added medical service and global cost of the solution.

There is, however, no absolute truth and no universal solution. The care sometimes escapes the statistics, and the medicine to science. There are two aspects of this. The first is given by precision medicine: individual discrepancies during a pathological state explain that we cannot only take into account statistics of a population to treat an individual, but try to refine its biological profile to adapt the therapeutics. The second aspect is well known to physicians and experienced by patients: medicine is not only a science but also an art, subtly delivered with personal, cultural, ethnologic, and religious considerations.

This anthropological vision of medicine will remain essential. Ancestral practice throughout the world, sometimes largely widespread like Ayurvedic or traditional Chinese medicine, should be taken into account as long as they fulfill the first mission of the care: to relieve the individual of his suffering.

At a time when old ideological divides must be forgotten in favor of efficient reforms, many health professionals want to move beyond a compartmentalized system to provide more effective, appropriate, and personalized care. They partner with entrepreneurs, researchers, and developers to make them understand our issues and reinvent the medicine we should be practicing in the twenty-first century, that is to say with our age-old ethical codes and the audacity to believe that technological progress will improve prevention, care, the patient's journey, and the ambulatory shift.

The caregivers' work is going to change, but they won't lose their jobs: whether clinical examination or a gesture of empathy, nothing will replace a human hand placed on another body. But the digital solutions must allow a simplification of the processes, an empowerment of the user, and a fluidity among too heavy systems.

We must also remain attached to the concept of health democracy, not by ideology, but because in everyday life it is nowadays normal in modern societies that everyone can be informed about his state of health, consciously make a decision concerning him, and freely choose to seek medical advice or support.

Finally, it is also a matter of social link. A healthy society is a society where many individuals are able, at a given moment, to make every effort to maintain one's well-being. This is the meaning of both prevention and care. A modern and efficient medicine is a practice that not only ensures good health but also restores the link, allowing individuals to regain autonomy in the face of illness or disability.

References

1. Topol E. The creative destruction of medicine. New York: Basic Books; 2011.
2. Schumpeter JA. Capitalism, socialism and democracy. New York: Harper; 1975.
3. Blue Button | HealthIT.gov. 2018.
4. Bergen M. Google is training machines to predict when a patient will die. Bloomberg.com. 2018. https://www.bloomberg.com/news/articles/2018-06-18/google-is-training-machines-to-predict-when-a-patient-will-die.
5. Rajkomar A, et al. Scalable and accurate deep learning with electronic health records. npj Dig Med. 2018;1(1):18.
6. Vivienne C. Système de santé. Aux grands mots les grands remèdes. https://itunes.apple.com/fr/book/syst%C3%A8me-de-sant%C3%A9/id1201265724?mt=11.
7. Macmillan A. La biopolitique et le dressage des populations. Cult Conflit. 2010;78:39–53.
8. @LeTelegramme. Déserts médicaux. "Les patients s'inquiètent". 2017.
9. Rasse P. La rencontre des mondes. Paris: Armand Colin; 2006.
10. Winkin Y. Anthropologie de la communication de la théorie au terrain, Points Essais. Paris: Éd du Seuil; 2001. p. 332.

Index

© Springer Nature Switzerland AG 2019
A. André (ed.), *Digital Medicine*, Health Informatics,
https://doi.org/10.1007/978-3-319-98216-8

MIX
Papier aus verantwortungsvollen Quellen
Paper from responsible sources
FSC® C105338

FSC
www.fsc.org

If you have any concerns about our products,
you can contact us on
ProductSafety@springernature.com

In case Publisher is established outside the EU,
the EU authorized representative is:
Springer Nature Customer Service Center GmbH
Europaplatz 3, 69115 Heidelberg, Germany

Printed by Libri Plureos GmbH
in Hamburg, Germany